MW00985045

ANCESTORS

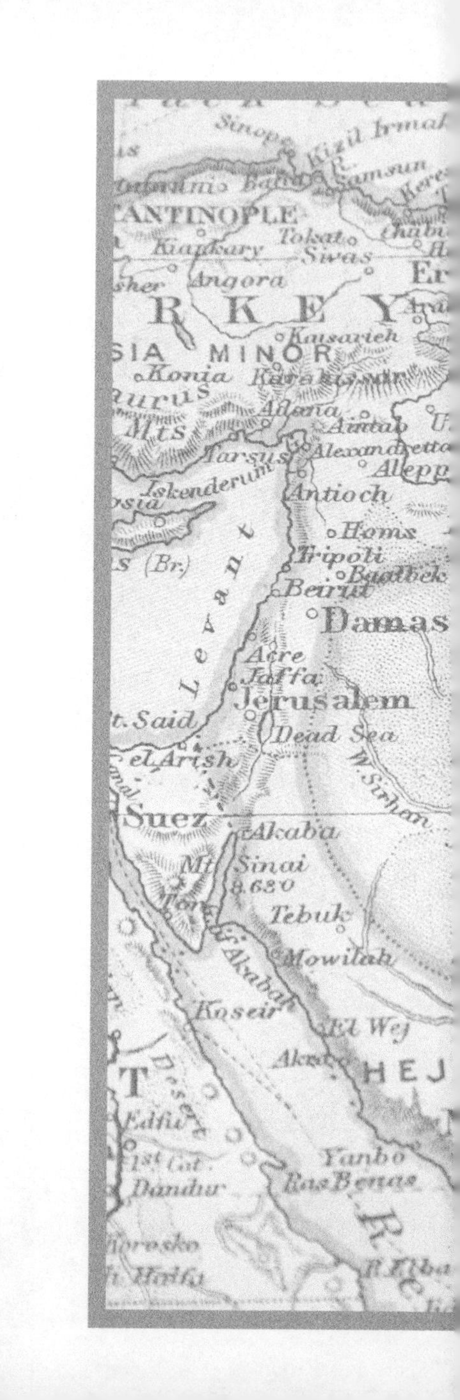

Sinope
Kizil Irmak
Samsun
Tokat
Kiankary
Swas
Angora
ASIA MINOR
Kaisarieh
Konia
Taurus
Mts
Adana
Aintab
Alexandretta
Tarsus
Iskenderun
Antioch
Homs
Tripoli
Baalbek
Beirut
Levant
(Br.)
Acre
Jaffa
Jerusalem
Dead Sea
el Arish
W. Sirhan
Suez
Akaba
Mt Sinai
8,620
Tebuk
Mowilah
Koseir
El Wej
Yanbo
Ras Benas

ANCESTORS

. . .

IDENTITY and DNA in the LEVANT

. . .

PIERRE ZALLOUA

RANDOM HOUSE
NEW YORK

Random House
An imprint and division of
Penguin Random House LLC
1745 Broadway, New York, NY 10019
randomhousebooks.com
penguinrandomhouse.com

LIBRARY OF CONGRESS CATALOGING-IN-PUBLICATION DATA
Names: Zalloua, Pierre A., author.
Title: Ancestors / Pierre Zalloua
Description: First edition. | New York, NY: Random House, 2025. |
Includes bibliographical references.
Identifiers: LCCN 2024046240 | ISBN 9780593730904 (hardcover) |
ISBN 9780593730911 (ebook)
Subjects: LCSH: Ethnic groups—Middle East. | Sociogenomics—Middle East. |
Genetic genealogy. | Human genetics. | Middle East—Civilization.
Classification: LCC DS58 .Z35 2025 | DDC 305.800956—dc23/eng/20241118
LC record available at lccn.loc.gov/2024046240

Printed in the United States of America on acid-free paper

1 2 3 4 5 6 7 8 9

FIRST EDITION

Book design by Barbara M. Bachman

Title-page map by Chad McDermott from Adobe Stock

Illustrations by Pierre Zalloua

The authorized representative in the EU for product safety and compliance is Penguin Random House Ireland, Morrison Chambers, 32 Nassau Street, Dublin D02 YH68, Ireland. https://eu-contact.penguin.ie.

To the memory of Louli

INTRODUCTION

A Few Things We Don't Quite Get About the Levant

By NASSIM NICHOLAS TALEB

Headdress for success—Nobody told the Arabs—Ivermectin and religious conviction—Bad news for Baden Baden—Mother Anatolia—Take away the State Department

SOME PEOPLE BELIEVE THAT the Levant is the end of the East and a portal to the West; others describe it as the end of the West and a portal to the East. Those in the first group tend to belong to the main branches of the Islamic faith, while those in the second belong to various Christian Levantine churches. Now, one might think that the two descriptions are equivalent: An intersection, after all, is an intersection. However, by the same mechanism that generates the so-called narcissism of small differences, not only are these two statements not equivalent, but they are, in practice, contradictory. It even took a civil war for the Lebanese to understand this fallacy.

Pierre Zalloua is a population geneticist. For a geneticist, there are (almost) no distinct categories, just continuous gradations. Many nationalists and racemongers thought that the new science of genetics would vindicate them and are currently hit by quite a surprise. Unlike languages, which tend to be discrete, origins are necessarily mixed: West and South Eurasian and North African populations are largely combinations of a handful of base populations, though in various proportions. Modern ethnicities result largely from dynamic admixtures of these same base groups over the past few millennia (though with different weights). The incoherence of nationalism lies in the belief that such mixing must stop on its last iteration; they've somehow reached their destination.

For instance, the French and the Germans live in divergent, even historically antagonistic national narratives; they speak mutually unintelligible languages while being, for the most part (excluding the South of France), genetically indistinguishable. Entire historical accounts have aimed at creating a racial separation in a self-feeding way. We will further discuss how, counter to the "European" narrative, ancient Greeks and the northern Levantines are also the same population, separated by language. Counterintuitively, the beauty of genetics resides in that it does not establish races but, rather, destroys long-held concepts.

Accordingly, this book removes some of the clichés related to the Levant that have accumulated over the past couple of centuries.

So the healthy way to think of the Levant is neither East nor West and, better, above such dichotomy; both its location on the Mediterranean and its proximity to the Caucasus and Arabia (though separated by a desert) are highly deceiving.

The area has inflamed the Western imagination for a long

time, partially explained by the technology and cultural transfers that took place over three millennia. For Westerners, there has always been an aura of holiness and mystery, not just from its origination of Christianity but also for the various deep Gnostic creeds built in the Levant, the numerous religions still buried there with secret beliefs that require a complicated initiation, often hidden under Islam or even, as we now know from the Gnostic Gospels, Christianity.

THE SUPERFICIALITY OF CULTURAL MARKERS

Let me first discuss how the notion of East versus West is highly constructed—and very poorly so. Consider the modern symbols that provoke such strong emotions. We'll start with the headscarf, indicative of a certain confession, with the French roiling over its non-republican significance. Well, the Islamic practice either originated from the Orthodox Church or was a common trait of the times across cultures—just look at old Russian babushkas. A noticeable trait in Renaissance paintings: Medieval Europe had stern sumptuary rules, with women dressed in black (and their heads covered) while men were allowed sartorial flamboyance.

But somehow, it recently spread as a cultural marker—in my own Levantine childhood, I never saw my Greek Orthodox grandmother without a head covering, while Muslim women (particularly in rural areas) were often bareheaded. Now consider the prayer style: ditto. The Orthodox performed the μετάνοια, with the head touching (even hitting) the ground during prayer, a symbol of both physical and spiritual prostration, a practice that was prevalent at the birth of Islam. Finally, consider architecture: The dome is Byzantine, and the first Islamic places of worship were designed by Byzantine architects—it was natural, as nobody informed the Arabs that there was supposed to be an East-West dichotomy that came with a cluster of symbols.

Further, the East-West demarcation, until the late eighteenth century and the formation of the nation-state of Greece, was not along current lines. The "East" started at the border between the Ottoman and Hapsburg empires—Orthodoxy thrived under the turban, which it preferred to the tiara. So did the name Levant. The Levant Trading Company, which brought coffee to coffeehouses in London three centuries ago, was based in Smyrna, now İzmir, Türkiye. And it was not until Greece joined the European Union in 1981 that people stopped saying "I am going to Evropa," meaning Western Europe. For "Levant" is a French exonym meaning "East," referencing France, similar to how "Anatolia" refers to "East" from the perspective of the Attic mainland. Its Arabic name, "Bilad El Sham," is an exonym meaning "North." "Canaan" is the only endonym of which I am aware.

Over time, with the rise of monolinguistic nation-states after the dissolution of the Ottoman Empire, the designation "Levant" kept shrinking until it referred to an area that is linguistically coherent, that speaks a collection of dialects that are mutually understandable under the umbrella "Levantine Arabic," mapping to today's Syria, Lebanon, and the Holy Land, closest to ancient Canaan.

DOURA-EUROPOS AND THE ARROW OF HISTORY

Now let us return to that identity business. About a decade ago, I was privileged to visit an exhibit from Doura-Europos, a Syro-Mesopotamian frontier city of the Roman Empire, in the area currently known for the origination and dominance of ISIS/ISIL (the Islamic State of Iraq and Syria/Levant). A synagogue had biblical scenes painted, proving that as late as the fourth century, human representations were still present in some parts of Judaism. Further, the same room served as a place of worship for pagans, Christians, and Jews.

The first lesson, which was a main theme of my book *The Black Swan,* is our endemic misunderstanding of the dynamics of history. What I call the "retrospective distortion," affecting both the perception of the random character of events and the degree of differentiation of past designations: We flow modern distinctions such as Christian or Jew back to the past in a severely deforming anachronism. We also pathologically like to categorize, and underestimate past diversity.

The second lesson is that religious differentiation, culminating in the modern polarizations, is a rather late thing in history. So is its ancillary religious intolerance.

So the direct ancestors of the most intolerant people on earth, ISIS supporters, would put to shame the modern West with their tolerance. We have numerous accounts in Syro-Palestine of families and tribes straddling different religions—some clans switching between Christian Maronitism and Mountain Shiism (Harfoush), some between Druze and Maronites (Abillama), some between Sunni Islam and Christianity (the Shehabs). Somehow the deadline for conversion closed sometime in the late nineteenth century. Even Sunni Islam (supposed to be the most orthodox) was quite differentiated, as some Sufi branches (in areas under Ottoman rule) were not aware of the interdict against the consumption of alcohol.

The rise of connectivity (in sequence: newspapers, radio, television, then the internet) had the perverse effect of making religions less regional, more centralized. For instance, Maronites used to swap religion with a Shiite neighbor in case of the realization of a pledge or vow—the practice has now stopped. My theory is that many religions stayed hidden under the cover of Sufi, Shiite, Alevi, Alawi, and Ezidi, and we have been losing such diversity.

So today we tend to observe in the Levant greater polarization around religion, with separation of groups along sectarian lines, the main ones being: Sunnis, Shiites, Druze, Maronites, Greek Orthodox, Monophysites or "Syriac Orthodox," Armenian Orthodox (the last two being incompatible and not in communion with the earlier ones), and Jews. But this, as we showed from Doura-Europos, is not intrinsic to religion: It is not necessarily a property of religion to produce polarization with clustered beliefs.

Yet if Levantines were in the past (somewhat) wise about religion, they were still polarized, but for other, generally silly things. It's a human trait to cluster and form networks with irrational and unexplainable clusters of viewpoints—as one buys a collection of creeds as a single block. At the time of this writing, those who vote for Donald Trump believe in the therapeutic benefits of Ivermectin, an equine dewormer, while his opponents prefer to rely on vaccines; now try to see any valid, uncontrived connection between therapeutic choices and political beliefs. So when Doura-Europos was a bustling city, with religion a noncentral marker, Levantines weren't textbook angels. There was almost always a severe brand of sectarian tension. The distinction was at some point between the blues and the greens, people rooting for different teams in hippodrome races, with some political and sometimes theological correlation. Later, after the rise of Islam, a new division was formed between Qaysi and Yamani, based on imagined ancestry, which cut across religious groups: There were Muslim, Druze, and even Christian groups on both sides.

THAT EUROPE THING

Some news for you, uncovered by Pierre Zalloua and colleagues (including yours truly): Where are you most likely to find people who are genetically closest to the ancient Greeks, those who have taken hold of Western imagination and fueled theories of cultural ascendance? Well, it's not in Athens, and even less so in the capitals of "Europe."

The West—that is, to simplify for the purpose of this foreword, Northern Europe, the land of butter—has spent some time trying to give itself its own letters of nobility by associating with the Greeks. This includes attributing some qualities as unique to the Hellenes—by assuming that the absence of evidence is evidence of absence and that if we don't have texts on a subject that precede the Greeks, then they must have invented it. This culminated with the great racial awakening of Europe emerging from the Middle Ages, which placed Indo-European and Semitic into separate categories, essentially language groups that became associated with race. Later, a German could claim to be vicariously, through their ancestors, an active partaker in the origin of "Western Civilization."

Our discovery will certainly irritate, perhaps infuriate, some neo-Nazi with a tattooed skull doing combat training in a clandestine camp near Baden Baden—or, even better yet, some professor of classics in a German university. For we found that the closest genetics to ancient Greeks' are those of people from the northern Levant. We sampled three groups from northern Lebanon near the Syrian border: Muslims from Dinniyeh, Maronites from Zgharta/Ehden, and Greek Orthodox from the Koura Valley. They proved to be genetically closer to the ancient Greeks than a random person in today's mainland Hellas. Is this from gene swapping with Greek sailors and Byzantine armies? No, as we could time the admixtures: The origin is very ancient. It just happens that both ancient Greeks and ancient Levantines, particularly those in the north, originate from the same source population in ancient Anatolia. For the distinction Indo-European versus Semitic is merely linguistic, not racial, but even scientists spent two centuries under the delusion that it was an unbridgeable genetic fissure. Languages move faster than people. As I argued in *Skin in the Game,* genes follow a majority rule, slow to disperse among neighboring populations, while languages (and religions) follow a minority rule, and can spread nonlinearly, like wildfire. If ten people whose mother tongue is not English sat in a company meeting with a solely English speaker, English will be the spoken language, which explains how the elite language becomes rapidly the norm. Often, as in Türkiye, Morocco, and Egypt, conquered populations change language rapidly, deceiving themselves about a national origin. Even Ernest Renan, who was obsessed with the racial superiority of the West, at some point defined "Semitic" as a merely linguistic category.* An easy way to see that the Phoenicians and Greeks were almost the same branch: The dominant paternal haplogroup for both is the Anatolian J2a (Zalloua will make the notion of haplogroup clearer in Chapter 9).

The Anatolian connection doesn't require genetics; it is actually visible to anyone with acceptable vision (or a competent optician) who has taken a road trip across the eastern Mediterranean. If you can't visually differentiate between a western Turk, an island Greek, and a northern Levantine, it is clearly because they share the same ancient ancestry. Further, (1) What is called "Arabic music" in the Levant is actually an Anatolian style shared across the entire zone from Crete to the foothills of the Zagros Mountains—the tones are the same, although the words change. (2) The dances: Lebanese and Syro-Palestinian dabkes are similar to those found in Turkey (halay) and the Greek world (καλαματιανός, χασάπικο, and συρτάκι). (3) The traditional Lebanese mountain garb, specifically loose-fitting trousers with tapered ankles, resembles the Greek βράκα and the Turkish şalvar,

* *Vie de Jésus,* 1863.

and perhaps has its origins in the Persian shalwar. (Pants, often misperceived as a Western thing, actually originated in Persia and Anatolia.)

THAT BAD GREEK OF SYRIA

Another fact that may irritate the "West versus East" crowd: The Levant was the larger contributor to "Western Civilization" directly, that is the Greco-Roman corpus during the Hellenistic era, the ten centuries extending from the arrival of Alexander in the fourth century BCE to the seventh century CE, with the Arab invasion. Just as I am now writing in English without being a Northern European and liking butter (and porridge), Levantines wrote in Greek while being diglossic (Greek and Aramaic), or even triglossic (Greek, Aramaic, and the local Canaanite dialect). There was even quadriglossia in Berytus, present-day Beirut, as the law school taught in Latin, something that horrified the purist Libanius Antiochus. (Ironically, I recall once insulting someone by telling him, "We spoke Latin before you," as, in addition, the author Ammianus Marcellinus was, as his name indicates, from Ammia, present-day Amioun, my ancestral village and place of residence.) Many historians miss the link, focusing on the East-West transfer of knowledge via the translation program of the Abbasids' House of Wisdom. Let us not forget that the New Testament was written in Antioch in what Nietzsche called the "bad Greek of Syria."

In a funny mix-up, the racemonger Charles Murray

wrote a book proclaiming the superiority of Western Civilization (and the "European" race) by listing its contributions, not realizing that most ancient names were Levantine.

So below is a short list of Greek-language Levantine authors and thinkers:

First, six Levantine scholarchs—that is, heads of Plato's Academy: Diogenes of Phoenicia, Hermias of Phoenicia, Syrianus, Marinus of Neapolis, Isidorus of Gaza, and Iamblichus of Apamea.

Next, the scholars Lucian of Samosata, Posidonius of Apamea, Numenius of Apamea, Zeno of Citium (originator of Stoicism), Zeno of Sidon, Philostratus, Philo of Byblos, Aeneas of Gaza, Libanius Antiochus, Zacharias Scholasticus, Boethus of Sidon, Apollonius of Tyre, Procopius of Gaza, Damascius, Apollodorus of Damascus, Domninus of Larissa, Timotheus of Gaza, Nicomachus of Gerasa, Ammianus Marcellinus, Antipater of Sidon, Antipater of Tyre, Marcus Valerius Probus (Probus of Berytus), Vindonius Anatolius Berytius, Dorotheus of Sidon, Hermippus of Berytus, Sopater of Apamea, Procopius of Caesarea, Lucius Julius Gainius Fabius Agrippa of Apamea, Antiochus of Ascalon, Apollodorus of Seleucia, and Cassius Maximus Tyrius (Maximus of Tyre).

Add the jurists Papinian (Aemilius Papinianus) and Julius Paulus Prudentissimus.

Finally, we count the theologians, John Chrysostom, John of Damascus, Ananias of Damascus, Alcibiades of Apamea, Alexander of Apamea, not including a certain number of Catholic popes.

THE MOST STABLE REGION OF THE WORLD

Another counterintuitive fact: The Levant today appears to be an unstable part of the world, filled with warfare and intractable conflicts. This may lead people to overlook the fact that between 1860 and 1948 (or, more generally, 1967, and, for Lebanon, 1975), the Levant was the most stable part of the world (in spite of an episode of rural famine in the early twentieth century). It attracted a large number of what the French called "Levantins"—French and Italian merchant families in search of stability who settled in the Ottoman Empire, some of whom moved farther south after its dissolution. There were bankers, doctors, dentists, piano teachers, and even French grammar specialists among these fortune seekers. Additionally, the Levant was the main destination for Anatolian Armenians following their massacre in 1915. Remember, during that period, Europe was consumed with warfare, from the Franco-Prussian War to the Second World War, which completely spared the Levant. The area benefited from not being on the radar of the U.S. State Department, an institution that wreaks havoc when trying to "improve conditions"—usually unsolicited—yet remains unaware of its track record. It is my hope that we will revert to that situation at some point in this century.

CONTENTS

INTRODUCTION BY
NASSIM NICHOLAS TALEB *vii*

PREFACE *xxi*

PART I. **ANCESTRIES AND IDENTITIES**

CHAPTER 1: Origins and Identities *3*
CHAPTER 2: Ancestry or Heritage? *16*

PART II. **FROM AFRICA . . .**

CHAPTER 3: From Africa to the Levant *29*
CHAPTER 4: The DNA Trail *37*
CHAPTER 5: *Homo sapiens* Meet Neanderthals
in the Levant *49*

PART III. **THE EARLY SETTLEMENTS**

CHAPTER 6: Our Early Ancestors in the Levant *61*
CHAPTER 7: From Anatolia and the Zagros
to the Levant *73*
CHAPTER 8: The Levant in the Neolithic Period *82*

PART IV. A COMPLEX GENETIC MAKEUP

CHAPTER 9: Population Expansions 89
CHAPTER 10: What Is an Indigenous Population? 105

PART V. HUMAN MOBILITY AND CULTURES

CHAPTER 11: The Early Dynasties and Empires 125
CHAPTER 12: The Early Tribes 137

PART VI. THE PHOENICIANS AND THEIR ALPHABET

CHAPTER 13: The Phoenicians 155
CHAPTER 14: The First Alphabet 168

PART VII. RELIGION AND THE MAKEUP OF THE MODERN LEVANT

CHAPTER 15: A Complex Narrative 183
CHAPTER 16: The Religions That Shaped the Modern Levant 197

ACKNOWLEDGMENTS 217

NOTES 221

PREFACE

Genetic ancestry testing—now widely available through private companies like 23andMe, Family TreeDNA, Ancestry.com, MyHeritage, and many others—has had a profound impact not only on the scientific community but on the general public as well. Scientists and nonscientists alike are rightfully concerned about how this data might be applied to the question of people's origins, and even their identities. In the same issue of a scientific journal, articles related to human population genetics and phylogeny are read and downloaded at least twice as often as the ones on medical genetics or other sciences.

On an almost weekly basis, major news outlets run stories on a new finding about ancient genetic relationships or phylogeography: Our immunity is shaped by our Neanderthal ancestry, the Americas were populated earlier than previously thought, DNA recovered from Egyptian mummies offers important clues about Egypt's history, DNA studies of the Black Death bacteria and the people who died from it explain why this pathogen was so deadly, and many similar stories that are significantly contributing to our understanding of human history and evolution.

Around the world, genetic ancestry testing kits are now marketed as ideal Christmas or birthday gifts. As these tests have become directly available to the nonscientist consumer, results and interpretations based on the provider's varying testing criteria and data availability have become complicated and challenging, often leaving the consumer confused. The appeal of genetic ancestry and its potential for unraveling "hidden mysteries" in our genomes has also led to an oversimplification and extrapolation (distortion, even, in some instances) of the results obtained from genetic studies that is unnerving and dangerous.

Origins, ethnicities, heritage, identities, and even race are being used interchangeably with genetic ancestry, with little or no attention being given to the complexities and dynamics that underlie these concepts. For example, determining whether an individual has Aboriginal, Native American, Māori, or Swahili heritage based on a DNA test can be very misleading. Before their first arrival in the South Pacific, the Americas, or the East African coast, populations have been mixing and continue to mix with Eurasian and other populations, making the interpretation of DNA ancestry testing very challenging. Restricting someone's heritage to some genetic patterns or signatures of their genome is disregarding the far more important cultural attributes (rituals, language, habits, diet, beliefs, etc.) that constitute the core of someone's heritage.

Origins, ethnicities, heritage, identities, and race are mostly social determinants or attributes and should not be debated or decided upon through the narrow lens of population genetics. To understand what any indigenous population is, one can't ignore genetics, but one must also look beyond it.

. . .

Since the dawn of time, humans have been migrating from one habitat to another—and sometimes back again. Our ancestors migrated for different reasons, but mostly it was in search of better habitats to improve their livelihood. This has been the case for millennia, and it has only increased in frequency in recent centuries. All the while, communities have exchanged languages, tools, artifacts, cultural habits, and rituals—as well as their genetic information—with other populations. As modern humans went from the hunter-gatherer lifestyle and became more sedentary, they claimed and inhabited territories as small groups. They cultivated the land, and as they developed farming skills and increased in population size, they expanded. Geography, terrain topography, and climate dictated how each group evolved and what habits and rituals (farming, hunting, housing, etc.) they had to acquire to survive best in their environments, and these became identifying signatures for these groups. Some groups remained relatively isolated, while others mixed and blended into larger groups. With time, these groups differentiated further by geographical boundaries and exposure to different and sometimes extreme environments, then morphed into distinct "cultures." Groups with distinct cultures evolved from other distinct cultural groups, and the geographical boundaries led to additional differentiations, as the groups became more diverse in languages *and genomes*.

With this in mind, we can see how the concept of indigeneity, whether at the individual level or the level of an entire population, is seductive but infinitely complex. Simplifying it for the sake of making sense of who we are, and where we came from, is not only erroneous but can be seriously misleading. Indigenous, or *ab origine,* populations have been historically and anthropologically categorized using various attributes such as geographical location, physical appearance, and language. Within their geographical boundaries and due to cultural practices and isolation, populations eventually acquire distinct genetic patterns. But assuming that indigeneity is based on genetic uniqueness is simply incorrect, and a misappropriation of population and evolutionary genetics. Certainly the more isolated the population is, the more it is correlated with specific genetic patterns that may over time become unique ("fixed," as geneticists would say) to that population. Isolation, however, has never been a condition favored by humans; human nature is rather more exploratory and favors interaction. While a few, indeed very few, populations remain isolated and have been isolated for a long time, the overwhelming majority have not.

Humans are some of the most sociable creatures on this planet, and none of the barriers—geography, genetics, language, and others—stood in their way of interacting with each other. Well before the establishment of any social structures or recognizable cultures, modern humans have been interacting and mixing not only with other modern human groups in different geographical regions, near and far, but even with Neanderthals and Denisovans, who in the not-so-distant past were considered to be different species. Currently, the various populations across the globe are in fact a mixture of earlier genetically and, if I can use the term, *culturally* more distinct populations that were themselves derived from populations that were even more genetically and culturally distinct.

I was one of the main researchers involved with the Genographic Project, a study conducted and funded by the National Geographic Society, IBM, and the Waitt Family Foundation to investigate human genetic diversity across the globe. I focused on populations in the eastern Mediterranean region and North Africa. When my colleagues asked me how many indigenous populations I could collect DNA samples from, the question took me by surprise, and I had no idea how I'd go about answering it. I wondered whether I could call myself indigenous to the Levant, and what the term really meant for the populations I was investigating. Would *they* call themselves indigenous? The thought of being indigenous had never crossed my mind before my work on this project. I had always thought that the term had no genetic basis, and that it was a cultural or a social attribute, a term used to designate a group of people who occupied a virgin land before anyone else claimed it from them. An "indigenous" group, seeking better opportunities, could have migrated from one or more ancestral groups living in different geographical areas. Would that make the mother group also indigenous in the newfound land, since they share the same genetic stock? Or is it the time since the split that makes populations indigenous? That is, the older the split, the more "indigenous" they are to the region? The further I explored this line of questions, the more I realized the category of indigeneity was troublesome and elusive.

. . .

DNA analysis has become one of the most useful tools for establishing, with a high degree of certainty, how and when people and populations moved and with whom they exchanged genetic material. Recent DNA findings have had a major impact on the fields of history, anthropology, and archaeology, and these findings are being used to rewrite major parts of our historical journey beginning in Africa more than 50,000 years ago. As one might imagine, integrating recent DNA discoveries into our common and long-accepted knowledge of our past has not been easy or without controversy and serious challenges.

In some instances, DNA findings propose novel, and sometimes shocking, models of human mobility and history. For example, results from a combination of ancient and modern DNA studies have shown that about 90 percent of the earliest inhabitants of England derive their genetic ancestry from the "Beaker" (in reference to their drinking vessel) people who moved onto the island, most likely from Spain and Portugal, around 4,500 years ago; they replaced the early farmers of continental Europe that started to settle the island several millennia prior and of whom 10 percent remained. New findings are now suggesting that the Americas were first inhabited much earlier than 14,000 years ago after people crossed a land passage from Siberia to Alaska. These findings point to several migratory waves that may have brought people from the old continent to America starting 22,000 years ago, and some evidence even supports a maritime passage. DNA work that I am involved in suggests that modern Arabia and Yemen were mostly populated through back-migration from the Levant around 6,000 years ago, and that the ancient populations who lived in Arabia prior to this date are yet to be discovered.

We now have strong scientific evidence showing that our direct ancestors, *Homo sapiens,* interacted and mated with Neanderthals and Denisovans and that most of us today carry Neanderthal and/or Denisovan DNA in our genomes. These are some of the new models that are reshaping our understanding of the history of humanity, from the time when our common ancestors all lived in Africa to the present day, where we find ourselves dispersed across the globe.

. . .

The eastern shores of the Mediterranean Sea comprise the heart of the Levant, the land of Phoenicia. This land has witnessed human mobility since the first cultures were known to have existed there—a place where early empires flourished, only to later succumb and be replaced by others. Thousands of years ago, genes shaped cultures in this region, only for genes later to be shaped by culture, through perpetual persecutions by the dominant cultures that paraded throughout the Levantine landscape, very often leading to the isolation and sequestration of the less dominant cultures, reducing their genetic diversity.

The Levant is the birthplace of the first empires, the first written languages, the first codes of law, the first myths, the first religious institutions, and the first concept of "society."

It is a place where origins run deep and identities are complex, blurred, and politicized, and where, in the name of these identities, cultures that were the foundations of all modern Levantine cultures—the Arameans, Syriacs, Assyrians, and Samaritans, among others—are being squeezed into obscurity.

The Levant with its many cultures and complex histories seemed to me a perfect model to explore the question of identities and origins. It is a place that has been continuously inhabited since the first known cultures appeared. It has witnessed and continues to witness human mobility and admixture like nowhere else on the planet. It is a place where identities and cultures often intersect, and where identities get entangled with religious beliefs. The Levant is the site where Neanderthals and *Homo sapiens* first intersected and exchanged genetic material.

I have been investigating the genetic landscape of Levantine and regional populations for more than two decades. Most of the stories told in this book are based on my own scientific research into the modern and ancient populations of the Levant and Southwest Asia, imbued with personal experiences with Levantine cultures. Since I grew up in Lebanon, the question most often asked around me was, "Are we Phoenicians or Arabs?"

So who *are* the people of the Levant, and where did they come from? People who ask this question may have had a longing to reach into their past, to find some anchor allowing them to claim their heritage with some satisfaction, which very often they confuse with their identity or culture. I use the word *claim* because people in the Levant—and elsewhere—can sometimes feel that their culture, their heritage, or even their very identities are being appropriated, not always by force, by more dominant cultures. Dominant cultures perpetually promote social, cultural, and economic narratives that are politically or ideologically driven and that evolve into mainstream societies. These mainstream societies become the default haven for the less dominant cultures.

In the 1950s, Pan-Arabism emerged as a liberating movement from the Ottoman and Western colonizers and swept through most of Southwest Asia, including the Levant and North Africa. The goal of this movement was to unite the various Arab-speaking countries as a political force to face Western dominance. But this movement failed primarily because it was driven by a few political elites in a couple of nations using only the Arabic language as the main anchor for this transnational polity. It failed to valorize and absorb the various cultures of the many Arab-speaking communities. This movement peaked during the rule of the charismatic Egyptian president Gamal Abdel Nasser, but after his death it lost its momentum and morphed into a purely political, nonsecular movement, promoting a national identity narrative rather than the rich cultural denominator that many of these communities share. Such narratives are adding to the confusion of how people perceive or feel about their own identities and cultures.

. . .

I am writing this book to share my perspective on identities and cultures. Identities are forged from cultures; they get adopted not by the group or the population but by the individual. Identities evolve, mature, grow, and become the lifetime property of the individuals carrying them. Cultures, on the other hand, are the basket of activities and actions performed by the group or by a population with distinctive behaviors through space and time and over many generations. They evolve with time, mature, and get collectively transmitted through generations and engender a population's heritage. Cultures that have been in the making for centuries or millennia, and identities that are personal and ever-evolving properties of the individual, cannot and should not be reduced to a technical summary produced by a genetic test.

Throughout this book, I weave personal stories with scientific insights and historical context to explore the concepts of indigeneity, population origin, culture, ethnicity, and race. This book highlights the impact of cultural habits, practices, and human adaptation on our genetic makeup and how these practices shaped the spread of the human species. It uses the Levant as an exploratory set to expose the richness and complexities of many of its cultures that have often been lost in historical translations or misappropriated by religious and political ideologies. It is written for those who want to learn about their own heritage and who are curious about how genes and cultures co-evolved, branched, diverged, disappeared, or matured, along our long, complex but fascinating human journey since the first *Homo sapiens* left the African continent hundreds of thousands of years ago.

. . .

This book is structured in seven parts. Part I provides a context for the discussion of heritage, ethnicities, and identities. It gives a cultural perspective on identities in the Levant against the backdrop of the widely used heritage and ancestry testing that is oversimplifying stories of origins. Part II describes the early movements of modern humans from Africa and their early settlements in Southwest Asia and the Levant. It highlights the impact of the climatic fluctuations on the migrations and mobility of the early Levantine inhabitants. It discusses the transition of the early inhabitants from hunter-gatherers to farmers, as well as their interaction with the Neanderthals, and it characterizes the various events that shaped their movements in the Levant and Arabia. Part III highlights major cultural developments within the Levant and Mesopotamia. It examines the creation of early complex societies and the first city-states and provides a chronological narrative of the various populations and cultures that have influenced Southwest Asia.

Part IV provides a context in which genetic findings are correlated or de-correlated with archaeological and/or historical evidence. This part highlights the early inhabitants of the Levant and their likely origins. Part V describes the evolving dynamics and the first cultures and empires, and the tribes behind their rise. Part VI is dedicated to the Phoenicians of the Levant and their alphabet. It describes how, via the discovery of genetic signatures, a culture that may have been long forgotten or overshadowed through misappropriation and misrepresentation of historical and archaeological events can be brought to light. Part VII addresses the impact of major cultural adaptations, primarily the adoption of religions and the resulting religious wars, on the genetic pool of the modern inhabitants of the Levant. It describes the genetic impact of the Crusaders and other religiously motivated mobilities on the modern inhabitants. It describes how religious habits can lead to significant changes in the genetic structure of a population.

PART I

ANCESTRIES and IDENTITIES

CHAPTER I

Origins and Identities

When people leave their homelands, do they leave their identity? Do they need to acquire a new one?

WHY DO PEOPLE MOVE? How do they adapt? And how do they identify with one another around the globe? When people leave the place of their birth, when they leave the neighborhood where they grew up, some take with them images of olive orchards, mountains of pine and cedar trees, and a soothing smell of the land. Personally, my first approach to characterizing identity is a nostalgic one. An identity is a basket of memories and collectibles, and wherever you go, you take them with you, and you keep on adding more and more to this basket. Identity is not a static concept; it is a state of perpetual evolution, shaped by events, exposure, and interactions. At an early age, you identify with faces that you see on a daily basis—your family, your friends, then the neighborhood—and the community. It is people and places, cultures, habits, and expressions with which one identifies. Communities are identifiable units that form the basis of cultural identities. Then you leave, like many of us do, and now there are new places, faces, habits, and expressions with which you may or may not initially identify. But your identity evolves, you adapt and/or adopt. The more you share with others the more you identify with them, and you collect new memories that continue to shape your identity.

I grew up on the eastern shores of the Mediterranean Sea, in the heart of the Levant, the land of Phoenicia. With the genetic information that I have gathered over the last decades on the Levantine populations, I felt compelled to write about the richness of these Levantine identities and cultures that are often lost in translation or stolen by religious and political ideologies. I have summarized in the pages that follow, not without difficulties, in a few thousand words, several thousand years of rich and complex Levantine history. In doing so, I wanted to give a historical perspective that can provide a context for the events that led to the major human mobilities that populated the Levant. But I also wanted to mix my knowledge of genetics with my own interpretations and share my own experiences of identities, cultures, and origins, using the Levant as my playing field. I wanted to free the Levantine identity from its historic hijackers. I wrote the following chapters for those who ask questions about their origins, heritage, and identity.

THE SEMITIC LANGUAGE AND ITS PEOPLE

Several years ago, a student asked me how to define a "Semitic population." I initially thought that this was a straightforward question that deserved a simple answer, so I told her that the word *Semitic* has two meanings, a biblical meaning and a linguistic one. All the descendants of

Shem (Sem), one of the three sons of Noah, are referred to as Semites, according to the Bible, hence the biblical definition. The linguistic definition states that Semites are all those who speak a Semitic language. My student seemed quite satisfied with my answer at the time, but the more I thought about my answer, the more perplexed I became, and now more than a decade later I am even more perplexed that I cannot answer this seemingly simple and straightforward question. The Semitic language designation refers to the same biblical figure Shem, and while the two definitions I gave my student share the same denominator, they are drastically different when translated effectively.

It was the German historian August Ludwig von Schlözer who first introduced the term *Semitic* in 1771 to describe a language family that was spoken by a group of people who occupied the Levant. Since then, the term has been used to designate both peoples and languages.

Linguists argue that the Semitic language originated in the northeastern Levant around 5,000 years ago, expanded east with the Akkadian culture into Mesopotamia, and then dispersed farther south into the rest of the Levant, to Arabia and from there to Ethiopia. There are many groups and populations living today, in total around half a billion, who speak a Semitic language. The Semitic language family is divided into several branches based on geographical distribution: East, West, South, Central, and Ethio Semitic. Except for Hebrew, most of the East and West Semitic languages are extinct, like Akkadian and Ugaritic, or nearly extinct, like Aramaic. The remaining include Arabic and the many languages and dialects spoken in Ethiopia and other parts of East Africa, like Amharic and Tigrinya.

Language dispersal, however, does not necessarily correlate with genetic dispersal. When I was recently asked the same question about Semitic populations during a presentation that I was giving on Levantine genetics, my answer was: Genes and languages seldom coincide, and defining populations based on linguistic attributes is the "apple" while defining populations based on genes is the "orange." Therefore, I am not sure whether all the populations that speak Semitic today share more genetic ancestry (or other attributes) with themselves or with non-Semitic speakers.

The term *anti-Semitism* is not related to our discussion here. This term was introduced by the German Wilhelm Marr in 1879 to label prejudice and discrimination against Jews and their Jewish faith.

ORIGINS AND IDENTITIES

Where does origin fit in one's identity?

When the great Levantine novelist Amin Maalouf was asked why he used the word *origins* instead of *roots* as his title for the 2004 book he wrote about his family and ancestors, his answer was: "*Roots* is not a word that exists in my vocabulary."

The concept of "roots" is not a human characteristic—it corresponds to a place, a site that is fixed in time and space. "Origin," on the other hand, while it relates to a beginning, is not only the beginning and cannot be confined to one space. People move, and with them they carry their origins deep inside.

One is born with a set of specific genes that makes one unique. Genetic studies and testing provide information about one's genetic heritage and genetic ancestry. However, there is no test, genetic or other, that defines or explains origin.

I was born in Lebanon, the Land of the Cedars, purple dye, and the Phoenician alphabet. When I took a genetic ancestry test, the results I was given stated that my maternal ancestors came from Western Europe. I conducted additional analyses and concluded that my maternal ancestors came with the Crusades around 1100 CE (Common Era) and my paternal ancestors (I'm less sure about them—they're a bit more complex) came from Central Asia or perhaps the Caucasus. So what does that make me? My DNA tells me that my ancestors arrived in Lebanon from the Caucasus or Central Asia, that I have Crusaders' genes, most likely French (from additional analyses), but what about my origins? I am neither French nor Central Asian. The Phoenician culture runs in my veins—they were my real ancestors and the ones I claim. Being Arab, Lebanese, English, French, Phoenician, or American has nothing to do with genetics.

While genetics confirms one's ancestry, which is undoubtedly part of one's identity, it does not verify one's identity.

How does one reconcile DNA ancestry with origins? I simply don't. I learned a lot of exciting information from my genetic ancestry testing. I now have many more stories to share about my ancestors and how I arrived at my birthplace. None of this information had any impact on how I feel about my origins.

If two monozygotic (identical) twins were separated at

birth and reared far apart (it is rare, but it happens), would they still share the same origins? If they were to take a genetic test, they would figure out their common genetic heritage. But does this really tell them about their origins? They will share the same genes, they may develop similar habits, and they will have the same genes imprinted (with their own on or off switch, irrespective of the environment). They will certainly look very much alike, but their environment and how they and their genes interact with it will shape many aspects of their behavior, their actions, and who they become as individuals. Studies that have been conducted on twins reared apart invariably show many similarities in behavior, but these similarities are only marginally significant, which is expected given the important role of the environment.

While genetic ancestry derives from allele frequencies in given geographical areas, those frequencies were established well before we were born. Origins are much more complex. They derive from a lifetime of experiences. They take their first form at birth and get molded with time.

FROM ORIGINS TO IDENTITIES

"Just like the ancient Greeks, my identity is linked to a mythology, which I know to be false and which nonetheless I venerate as if it were a bearer of truth."

Amin Maalouf is speaking not about genetic identity, which has a narrow and scientific meaning, but about cultural identity, which is a much broader term and takes precedence. Suppose a thirty-three-year-old woman finds out that the parents who raised her are not biologically hers, that she was conceived by donors. Will she change her identity after knowing her biological parents? This example is an oversimplification of the concept of identity. It is important to know one's genetic pedigree, but is it a *sine qua non* for one's identity?

Origin and identity are not simple concepts by any stretch, and I am not, nor do I claim to be, an expert on the topic. As a geneticist, however, I find the interest in DNA ancestry, and the appeal of applying it to origin and identity, fascinating. Many thriving ancestry companies sell genetic ancestry to the public, but my fascination lies in the innumerable non-geneticist "hobbyists" who have been amassing large amounts of data, at their own expense in many cases, about specific phylogeographical signatures. They have become true experts indeed. This phenomenon is rarely seen in other research fields. Today many non-geneticist experts, some very reliable, have more knowledge about the distribution of genetic signatures than do many geneticists. Geneticists often seek their help and ask for their data when it is possible for it to be shared.

While I'm fascinated by this phenomenon, I'm equally unnerved by the oversimplification of identity and origin that results from genetic ancestry and some of its interpretation. If anything, genetic ancestry and phylogenetic studies have debunked the concept of "ethnicity," a word that I have become averse to using. I prefer *ethnic background,* but I would certainly advocate using the word *culture,* which I think trumps all other terminologies. Genes do not change, origins migrate, and identities evolve!

WE NEVER CALL OURSELVES ARMENIANS

Lebanon today is home to more than 200,000 Armenians who were born in Lebanon to Armenian parents (whose grandparents had escaped the 1915 Armenian genocide by the Turks). If you were to ask them whether they consider themselves Armenian or Lebanese, what would they say? Without hesitation they would answer: *We are Lebanese of Armenian descent*. In fact, Armenians never call themselves Armenians—they prefer to be referred to as Hay, in reference to Hayastan, their native land, which is now called Armenia. Hay, or Haya, is the name of a deity that was mentioned in Sumerian inscriptions in Mesopotamia, and Hayk is the name of the legendary ruler of Armenia. Some deciphered Hittite inscriptions refer to the Hay people and their native land, Hayasa. In Hittite inscriptions, the suffix *-asa* or *-sa* is used to designate a place or a habitat.

Today Armenians are well represented in all branches of the Lebanese government as members of parliament and cabinet ministers. Lebanese Armenians are a fully integrated community, but they never lost their heritage (cultural and not genetic). Armenian food has become a fixture in Lebanon, and almost every Armenian in Lebanon speaks Armenian, even those who come from mixed marriages. Armenian culture was transplanted a great distance, over a century ago, during the most difficult time of its people's existence. They fled thousands of kilometers away to a place where a totally different language family is spoken (from Indo-European to Semitic). The decision made by these people as early as 1875 to leave their native land and migrate great distances and endure hardships across many unfriendly geographical and political boundaries to reach Lebanon was extremely courageous.

Despite being in a foreign land, Armenians managed not only to preserve their culture but to integrate and thrive even as they kept their cultural identity intact. Many factors may have led them to choose Lebanon, but their decision proved to be a successful one, as they fully integrated into their host population with their own culture. These factors are intrinsic not only to the immigrating culture but to the host community as well.

THE ARAB IDENTITY CONUNDRUM

The Arab League is composed of twenty-two countries, spread over a huge geographical area across more than one continent. Its more than 220 million people all share a common language, Arabic. The league was created in 1945 after the Second World War by seven founding members, later joined by the remaining fifteen. In the 1950s an "Arabist" nationalist movement was created to promote social progress and shore up unity among the people living in Arab countries. This movement was later to be conflated with what is called the "Arab identity" or "Arab ethnicity," and the illusion that this "ethnicity" or "identity" is equated with a Peninsular origin.

Setting genetics aside for the moment (which show tremendous variability), almost all of the twenty-two countries have their own distinct cultures. These cultures are so old, rich, and diverse in habits and rituals that it is perilous to lump them under a unitary "identity" that has no fundamental basis.

MINORITIES

Social, political, and religious ideologies should not hijack origins and identities. Defining identities or "ethnicities" based on ideologies, political considerations, or linguistic or religious beliefs effaces culture and creates the hazardous concept of "minorities." What a scandalous word to use when describing human beings! What does it really mean anyway? How can we equate a human to a number, a culture to a metric or a quantity?

The word *minority* sneaked into our lexicon to designate a subpopulation or a group of people smaller in number than the larger, "dominant" population in a given space (a country, for example), usually having a different heritage or culture. When you type the word *minority* in Google, the top search result you get is the Google Dictionary definition: "the smaller number or part, especially a number or part representing less than half of the whole." But the same search also gives you other results about "ethnic minority," which on further interrogation reveal another definition from Google Dictionary: "a group within a community which has different national or cultural traditions from the main population." Fantastic, I thought, this is a great definition that is "politically correct." It defines "ethnic minorities" as people with different cultural traditions. Scrolling down, you come across the Cambridge Dictionary website, which gives the following definition: "a group of people of a particular race or nationality living in a country or area where most people are from a different race or nationality." And just below this definition is a list of synonyms and related words like *alien, anti-Semitic, color prejudice, ethnic cleansing,* and *foreign,* among others.

Let us imagine a scenario where the main population—or the main "race," as per the third definition—with time and due to demographic changes (which are happening now across the globe, have happened before, and will continue to happen in the future) becomes the "minority." How are we to describe an ex-majority population that has become the minority? How will demographic changes affect nationalities and identities? Why can't we stick to the word *culture* and forgo *minority*? Why does it have to be a zero-sum concept? No one should be a minority in their own community or country. Some cultures are being replaced by others, and some are getting lost in the process of defining minorities and majorities. Losing cultures is a major concern that humanity must grapple with—we must try to save cultures from eroding into extinction, and stop calling them minorities.

But cultures do integrate better than numbers, and cultural diversity is our only guarantee against extremism and extremist ideologies. Cultural preservation is the safeguard of diversity. A truly diverse, multicultural society has no place for dominance, majority, or minority. In his 1996 book, *The Clash of Civilizations and the Remaking of World Order,* Samuel P. Huntington argues fervently against the tendency for a universal culture and warns of the perils of a monoculture or Western cultural hegemony. He lists one of the major causes of civilizational clashes as religion.

Unfortunately, religion, especially in the Levant, has dominated or even replaced all other attributes of one's identity. People identify with religious affiliations and feel more comfortable with religious identities, and that is exactly why nation building has failed miserably in many places across the Levant but also elsewhere. Perhaps this was the consequence of the perpetual wars that have been raging in the Levant for two millennia, where religion was their main denominator. Fighting in the name of religion means fighting against some beliefs, rituals, values, and ways of life and fighting others in the name of defending faith.

In Europe, after the religious wars that had raged for many centuries came to an end, people drifted away from religious identities. Europeans evolved together with their religion. In 1615 one of the brightest European minds, Galileo Galilei, was charged with heresy by the Catholic Church and sentenced to prison because his arguments about the earth's rotation around the sun contradicted the Holy Scripture. Today *heresy* is an archaic word in the annals of European history. In the Levant, however, cultures evolved around their religion, coalescing people of similar beliefs and cultures, perpetuating the fragmentation of an already fractured Levant.

Over lunch a former dean at a highly respected Ivy League college and I were discussing the Levant's never-ending political struggles when he asked me, "How will the religious minorities in the Levant fare in the next few decades?"

"With questions like this, they will certainly fare poorly," I said. There are many cultures in the Levant, some have similar faiths and others do not, but they are all Levantine and have been for a very long time.

CHAPTER 2

Ancestry or Heritage?

The concepts of "race" and "ethnicity" take attention away from the individual

DNA IS A STATIC, OBJECTIVE characteristic that describes the journey that our ancestors have taken since they became *Homo sapiens*. The study of DNA yields effective tools that can unravel interactions and admixtures among various populations around the globe and corroborate or contradict historical narratives related to population movements. It provides a certain probability of the populations to which our ancestors may have belonged.

While DNA signatures are probabilistic percentages, habits and cultures appear to be more certain. Still, adopting a culture, a habit, and a language does not make you ethnically distinct.

IS THERE SUCH A THING AS "RACE"?

The first time I had to really think about the concept of race was when I was filling out an application for admission to a U.S. university. I did not know which box to tick for "your race"—I had never thought about "my race" and how others perceive me. I had never been in a situation before that prompted me to think about "my race." I had to call my brother, who was already living in the United States. He told me to tick the box where it said "White/Caucasian." "We are considered white," he said.

Caucasian was another term that I had to grapple with for a while. Today when you type the words *White Caucasian* into Google, you get the following definition: "(no longer in technical use) of, relating to, or characteristic of one of the traditional racial divisions of humankind, marked by fair to dark skin, straight to tightly curled hair, and light to very dark eyes, and originally inhabiting Europe, parts of North Africa, western Asia, and India."

The concept of race began in the eighteenth century CE with the Swedish biologist Carolus Linnaeus, who, during his work on the classification of species, made distinctions among *Homo sapiens* based on continental divisions, proposing four subcategories of *Homo sapiens* in his *Systema Naturae,* published in 1735: *americanus, asiaticus, africanus,* and *europeanus.*

In 1824 the German physician and anthropologist Johann Friedrich Blumenbach, building on the work of Linnaeus, proposed in his book entitled *On the Natural Variety of Mankind* five families of *Homo sapiens* based on cranium measurements: Caucasian, Mongolian, Malayan, Ethiopian, and American. Earlier efforts by the French physician François Bernier and other European taxonomists documented *Homo sapiens* variations based on phenotypic (morphological) features and used the words *type, race,* or even *species* to distinguish between them.

The word *race* was transformed from a physical property "objectively" based on morphological attributes (phenotypes and not genotypes) into a category differentiating humankind exclusively based on "subjective" social determinism heavily influenced by socio-economic criteria. The outrageous use of this "new" concept of "race" reached its pinnacle when Hitler and followers preached eugenics and the existence of a "pure race" that was genetically superior.

I had no prior knowledge of the history of race, and when I ticked the box indicating that I am "White/Caucasian," that was the end of it for me. Then when I moved to the United States, I noticed how important the concept and perception of race was to people. Today race means a great deal to a great many people around the globe, but it means different things to different people.

TESTING FOR ANCESTRY?

Genetic testing services like 23andMe, AncestryDNA, and many others provide a great deal of information about an individual's ancestry, which can unravel fabulous, otherwise hidden stories about one's past. These tests are based on analyses that generate DNA patterns. These patterns are then correlated, with a high degree of certainty, to a group, a population, or even a geographical location. The larger the number of people from the groups or populations tested (i.e., the companies' databases), the greater the chance to find a pattern similar to that of a given individual who wishes to get tested. But if the number of individuals tested in each population (the database) is very small, correlations cannot be made successfully, and the results are not specific. Therefore, ancestry testing providers with the largest and most population-inclusive databases are more likely to provide more accurate results.

Defining which populations or groups to target for genetic ancestry information is always tricky. In places like the Levant, where population mixing runs high and has for quite some time, the number of individuals required to obtain a reasonably recognizable DNA pattern characterizing the various groups in a population needs to be quite large. Even when we consider only paternal or maternal lineages, where recombination (genetic mixing that occurs in the gametes of the parents prior to fertilization) is not a factor, a very large number of various genetic signatures can still be observed, making correlations between genetic and geographical ancestry very difficult, if not impossible.

INDIGENOUS POPULATIONS

When a colleague asked me how many indigenous populations I could identify in the Levant, I was surprised by the question and my answer was, "Indigenous as opposed to what, inhabitants?"

When the Americas were discovered (or rediscovered, to be precise) in 1492 by the Genovese navigator Cristoforo Colombo (Christopher Columbus)—five hundred years after the Icelandic Leif Eriksson—millions of people were living there. But those residents were not considered the first discoverers of the Americas in the Eurocentric colonial view of the world back then. They were referred to as the indigenous people of the Americas, or American Indians, or Native Americans. The term *indigenous* is derived from the Greek word *endo,* meaning "inside" or "within." This term evolved to *indu* in Latin, which gave the Latin word *indigena,* which means "native to a specific place." The term *Indian* was used, prior to Columbus, to designate the people of the East Indies (India and the entire Asian subcontinent); the European colonizers adopted it to designate the newly discovered people of the Americas (thinking they had landed in Asia). The peoples of the Americas never called themselves indigenous or Indians, obviously. They have inhabited the Americas for several millennia through several waves of migration. The same narrative holds true for Australia, the Pacific Islands, and a few other places.

An indigenous, or *ab origine,* population is the one that first occupied a geographical habitat. Today "First Inhabitants" is used to differentiate them from the colonizers.

In the Levant, the word *inhabitant* simply means *ab origine,* "indigenous" or "First Inhabitant." Being First Inhabitants, however, does not mean absolute genetic homogeneity or the absence of genetic variability. (I doubt if such a group exists.)

All populations are made of a collection of subpopulations or subgroups. Not all First Inhabitant subgroups are genetically homogeneous, and not all genetically homogeneous subgroups are First Inhabitants. Some groups within a highly admixed population may become more genetically homogeneous because of nonrandom mating (in which people with similar genetic traits, or signatures, mate), followed by extreme isolation, while some First Inhabitant groups may have been heavily genetically admixed prior to isolation. In a nutshell, all populations today are genetically admixed, and some more than others. As we will see in chapter 10, even some of the most remote "indigenous" groups of Africa have European genetic admixture.

ETHNIC POPULATIONS

In the introduction to one of the most (if not *the* most) cited books on ethnic studies, *Ethnic Groups and Boundaries* (1969), Fredrik Barth wrote: "Since culture is nothing but a way to describe human behavior, it would follow that there are discrete groups of people, i.e., ethnic units, to correspond to each culture." The word *ethnicity* is derived from the Greek word *ethnos,* meaning a group of people living together, a tribe, a group of foreigners, etc. Ethnicity has many definitions, but they invariably refer to a group of people who share the same culture and language, have a unique designation (etymology), and maintain self-preservation/propagation (biology or, to be exact, mating).

The notions of ethnicity became less rigid after luminaries like Barth wrote about the permeability of ethnicities, canceling the concept of genetic ethnicity but maintaining cultural boundaries. The anthropologist Michael Moerman went even further, arguing that objective measures in defining ethnicity are in most instances useless. He spent years studying the Lue, a relatively small population living in Thailand, China, and surrounds. In his 1965 article "Ethnic Identification in a Complex Civilization: Who Are the Lue?" he wrote, "Someone is a Lue by virtue of believing and calling himself Lue and of acting in ways that validate his Lueness."

Among the Lue, ethnicities are not rigid designations. They are dynamic, their characteristics are often not coterminous, and they change over time and across geographies. In other words, they are permeable. Nevertheless, boundaries among groups are maintained through a complex network based on a plethora of social interactions, in a manner that Barth eloquently described: "Ethnic distinctions do not depend on an absence of social interaction and acceptance, but are quite to the contrary often the very foundations on which embracing social systems are built. . . . Cultural differences can persist despite inter-ethnic contact and interdependence."

Designating ethnic populations based on attributes that include genes has many shortcomings. Populations that constitute genetic isolates—or even confined social entities for that matter, if they exist (I am not sure they do)—would have to be completely isolated after being derived from a very small number of progenitors. Gene and cultural flow from surrounding populations into these isolates (through mating/marriage over several generations) would have to be very limited due to geographical barriers and/or cultural habits that favor endogamy (mating/marrying within the same group or clan). These populations would have a high degree of homozygosity (many of their alleles are similar) across their genomes, a high level of inbreeding (mating among relatives), and a small effective population size (fewer than one hundred individuals as progenitors). If these populations exist, one will need to study many individuals not only from the population in question but also from all the neighboring populations to be able to demonstrate genetic isolation. There are currently more than six thousand classified populations in the world, and I am not sure if any of them fit the definition of a purely indigenous population.

THE GAME CHANGER: ANCIENT DNA

Technology has brought another dimension to the understanding of human populations. We can now successfully obtain and analyze DNA from ancient human, animal, and plant remains, something that was impossible only a decade ago. As a matter of fact, the new information made available from DNA work on ancient materials has required us to revisit a lot of the conclusions derived from older DNA work conducted on human mobility. Work on ancient DNA found mostly in teeth and bones in archaeological digs has been instrumental in identifying the various ancestral genetic components in First Inhabitant populations. The technology is permitting us to determine whether there has been population replacement or genetic discontinuity between ancient, indigenous, *ab origine* populations and modern ones.

A seminal article on European populations by David Reich and his colleagues demonstrates the power of this technology and underscores the importance of including studies of ancient DNA when investigating populations. Reich and his colleagues analyzed DNA from the ancient remains of four hundred individuals across Europe. They demonstrated that almost all (90 percent) of the "indigenous" Neolithic population of Britain was replaced 4,500 years ago by the Beaker people. The name derives from the distinctive bell-shaped drinking pots that these folks used—they are found in almost all the burial sites. The exact origin of the Beaker people remains poorly understood, but carbon dating evidence traces the oldest Beaker to the Iberian Peninsula. In any case, genetic differences between the two populations' DNA are large enough that we can clearly distinguish the early occupants of Britain (indigenous British) from the Beaker people, who show similarity to the island's modern inhabitants.

I had always thought that population replacement was an unnatural phenomenon, as it is very hard to replace an established population. Not even the Neanderthals and the Denisovans were totally replaced—today's evidence shows that there have been at least four waves of admixtures between them and modern humans. Reich's Beaker replacement theory, based on analyses of ancient DNA, shows that when several factors come into play concurrently (climate change, microbial diseases, arrival of competition), a population can succumb and nearly disappear no matter how large it is. But researchers after Reich could not demonstrate the replacement phenomenon in other communities across continental Europe. While the Beaker people left plentiful evidence of their existence, the levels of replacement in continental Europe were far less distinct. Perhaps in Britain, replacement resulted from the movement of both people and culture, whereas in the rest of Europe, it resulted essentially from cultural movement without major human mobility.

In our work on the Phoenicians, we have witnessed a phenomenon of population replacement on the island of Ibiza. We studied the mitochondrial DNA (DNA that is passed exclusively from mother to child) of the island's first settlers ("indigenous"), who came from western Phoenicia, most likely Cádiz, around the sixth century BCE (Before Common Era). We found none of their mitochondrial DNA present in Ibiza's modern population. The first settlers' entire female population has been replaced, first by a population that arrived from North Africa around the eighth century CE during the Islamic expansion into Europe, and then around 1200 CE with colonists from mainland Spain. The demise of the earlier settlers is attributed to multiple factors, including microbial disease and inbreeding resulting from isolation.

ANCESTRY OR GENETIC HERITAGE

Multiple cases like the two just cited highlight the importance of ancient DNA studies in understanding how populations are derived. Based on DNA, one can define a population genetically and can learn a great deal about its genetic ancestry. One can even define with a certain degree of accuracy how many genetic components (ancestries) make up a population and how and when these components were derived from earlier ancestral components.

Genetic ancestry can be quite revealing about the past of one individual or an entire population. Adding geography to genetic ancestry and deriving population split times can identify how and when populations arrived at their current location and determine their entire ancestry with a high degree of certainty, especially when ancient DNA information is used.

Genetic ancestry is a powerful tool that can unravel one's genetic heritage (as opposed to one's cultural heritage, which is a much broader term). Heritage is a beguiling concept that is better used when referring to populations than to individuals. But genetic heritage, a narrower concept, can be used at the individual level and can unravel fascinating and unsuspected stories about one's past. I know, with a high degree of certainty, that my genetic ancestry is mixed. My maternal lineage is European, derived from a Crusader—I could probably track down his other descendants if I wanted to. From my father's side, my ancestors came from Central Asia or somewhere around modern-day Armenia (the Caucasus). As more data become available, my family will be able to trace our ancestry with a high degree of certainty. Genetic ancestry when combined with geography can be extremely accurate and can corroborate or debunk many stories and histories about individuals as well as populations.

PART II

FROM AFRICA . . .

CHAPTER 3

From Africa to the Levant

Every known human journey can trace its origins to Africa

WHERE DO WE COME FROM?

Archaeologists long believed that *Homo sapiens,* anatomically modern humans, evolved in sub-Saharan Africa around 200,000 years ago. The oldest *Homo sapiens* skull, discovered in the Great Rift Valley in Ethiopia, was dated to about 195,000 years ago. The first archaeologically documented migration of *Homo sapiens* out of Africa was believed to have occurred between 200,000 and 150,000 years ago, but it was not successful, as modern humans could not adapt to the harsher, cooler environment that existed outside Africa at the time. Until very recently, multiple theories have been proposed about when and how *Homo sapiens* left sub-Saharan Africa, but none went back in date beyond 200,000 years ago.

In 1961 archaeologists were excavating a cave in Jebel Irhoud in western Morocco, North Africa, when they came upon a skull. Their discovery was long neglected, but in late 2017 a group of researchers took another look at it and made a startling discovery. It was actually a *Homo sapiens* skull, and it was 315,000 years old.

Then in 2019 another group, using novel dating technologies, reanalyzed a skull fragment from a human fossil that had been excavated in 1970 from Apidima Cave in southern Greece. They reported that it too belonged to *Homo sapiens,* and they dated it to at least 210,000 years ago. It is, to date, the oldest known modern human remnant ever found outside the African continent.

These discoveries put in serious doubt, maybe even shattered, all the long-accepted dates for the evolution of our species and our ancestors' first dispersal out of sub-Saharan Africa.

MULTIPLE WAVES OUT OF AFRICA AND BACK-MIGRATIONS

The Apidima Cave skull places *Homo sapiens* outside Africa around 210,000 years ago, suggesting that the first modern human migrants most likely left Africa before that date. But we have no other clear evidence, genetic or otherwise, that this migration, or any earlier ones that might have occurred, led to the successful and permanent establishment of modern human populations in any location outside Africa.

EARLY MIGRATIONS

It is widely accepted that *Homo sapiens* left the African continent (1) to search for better hunting grounds and (2) to escape a massive drought, one of many that have plagued the African continent since the Last Glacial Period (110,000 to 18,000 years ago), when the climate in Africa shifted from moderate to so hot and dry that it could have forced modern humans to leave their African habitats. They may have moved north through the Gate of Tears (Bab El Mandeb) in Yemen or through the Levantine Corridor (the Sinai). They were hunters and gatherers, and they spread in small groups throughout Southwest Asia.

Solid and relatively abundant archaeological evidence places modern humans in the Levant around 90,000 years ago and in Saudi Arabia around 85,000 years ago. Then, around 80,000 years ago, harsh weather conditions suddenly cooled most of Southwest Asia and the Levant, and modern humans disappeared from most of this region. Evidence of human habitation in the Levant and Arabia almost vanishes between 80,000 and 60,000 years ago, the date proposed for the subsequent and more successful migration waves, suggesting that these earlier migrants may have gone back to Africa to escape the cold weather. It is not clear what drove the sudden change in climate. What is certain, however, is that the change was abrupt, and it may have forced these foragers to either escape or face extinction.

THE EARLIEST HUMAN PRESENCE IN THE LEVANT

The two most often cited locations in the archaeological record that document the presence of modern humans prior to 60,000 years ago in the Levant are the Skhul and Qafzeh caves on the slopes of Mount Carmel in the Galilee. These sites contain *Homo sapiens* fossils dated to around 100,000 to

90,000 years ago. Other sites in the Levant, like the Tabun, Kebara, and Geula caves, document the contemporaneous presence of Neanderthals.

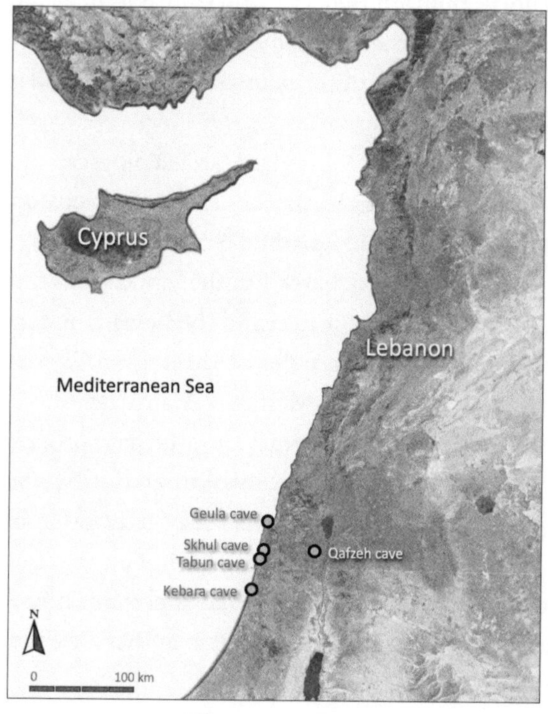

The ancient Levant

Modern humans and Neanderthals coexisted in a small geographical area within the same period of time, but there is no evidence that they concomitantly occupied the same habitats. However, the two groups may have interbred. Their coexistence was not destined to last long, as the sudden change in climate favored the Neanderthals, who were better adapted to cooler climates. *Homo sapiens* left the Levant to return to Africa through the Levantine Corridor in search of a warmer climate, while the Neanderthals stayed. Eventually they moved on and reached Europe and Asia.

While *Homo sapiens* first appeared as a distinct species in Africa, we are still debating where and how they departed from Africa. Multiple migrations must have occurred, some more successful than others, over a long period of time. *Homo sapiens* must have migrated back and forth before they finally achieved a sustained presence outside Africa. Recently uncovered archaeological and DNA evidence now supports the view of multiple back-and-forth migration waves between Africa and Eurasia, and in multiple locations. Once these hunter-gatherers were out of Africa, they did not stay in one place—they kept moving in search of better hunting grounds and more suitable climates. To date, their migration patterns and expansions remain, however, poorly understood.

Climate, sea levels, and terrain topography certainly played a major role in the way these early humans moved from one hunting ground to another. To date, very few sites in Southwest Asia present archaeological and paleontological evidence of modern human occupation dating to the period ranging from 60,000 to 18,000 years ago, when the ice of the Last Glacial Period started to melt. The cold weather that persistently dominated the Levant up until the Late Paleolithic (18,000 years ago) restricted movement of the early inhabitants of Southwest Asia and most likely confined it to specific shelters or refuges.

IN SMALL GROUPS

The early inhabitants of the Levant, as hunter-gatherers, most likely lived in small groups in the few geographical areas that supported wildlife and vegetation. Reconstructed climate maps based on pollen records and oxygen isotopes (O^{18}) taken from frozen or dried-out lakes and speleothems (stalactites and stalagmites) reveal the presence of geographical areas where the climate was optimal for the survival of plants and animals. The Levant's early inhabitants used these areas as refuges.

Climate maps that are based on pollen cores and oxygen isotopes consistently show a mostly arid Levant during this period. Generally the western Levant, the area along the Lebanese coast, was wetter than the eastern part, extending to the Jordan Rift Valley and Arabia, where precipitation was scarce.

During the Late Paleolithic (around 18,000 years ago), the Levant witnessed a major cultural shift marked by continuous technological advances in lithic (stone) toolmaking.

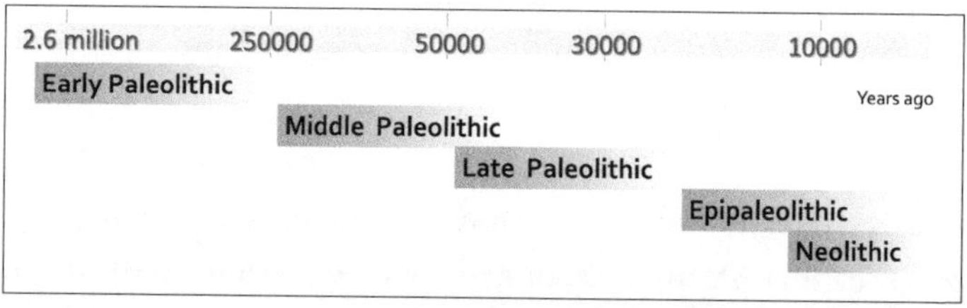

2.6 million 250000 50000 30000 10000
Years ago
Early Paleolithic
Middle Paleolithic
Late Paleolithic
Epipaleolithic
Neolithic

General archaeological timeline

During this shift, the crude hunting blades that had dominated the Late Middle Paleolithic (75,000 to 45,000 years ago) gave way to more refined stone tools and hunting blades. In toolmaking methodologies (transitional industry), this shift implies both continuity and evolution between two distinct periods, one dominated by Neanderthals and one by *Homo sapiens*. Recent DNA evidence shows that modern humans dwelling in the Levant are in fact the descendants of a Neanderthal and *Homo sapiens* mix. If these two types of humans did indeed interbreed, the Levant would have been the ideal place for it.

The presence of *Homo sapiens* in Southwest Asia 50,000 years ago is documented by certain fossils and tools found in caves throughout the Levant, Arabia, and Mesopotamia. The Ksar Akil cave in Lebanon is considered one of the most important Levantine sites documenting their presence. Most of these caves are found in a narrow strip of hospitable environment, delineated by the Mediterranean coast to the west and the Mount Lebanon chain with its variable topography to the east.

The use of animal bones and antler points in making tools and fine blades has been recorded throughout the Levant for the period between 23,000 and 12,000 years ago.

Modern humans started to live in large groups around 23,000 years ago, the date of the first recorded human settlement. During this period, the climate generally improved, with warming and moistening environments supporting vegetation, and modern humans expanded throughout the Levant and beyond. They developed new habits, including the collection of seeds and wild cereals. Grinding tools have been discovered in a few sites.

The period between 23,000 and 18,000 years ago marked

the beginning of community life, a major shift in human culture. For the first time in *Homo sapiens'* existence, the near-global hunter-gatherer lifestyle was replaced by a small community way of life. Throughout the Levant and Anatolia, early settlements appeared, constituting semi-sedentary habitats.

CHAPTER 4

The DNA Trail

Some tales are better narrated with DNA

EARLY PEOPLES MADE THEIR WAY to the Levant over multiple and distinct routes. Before I elaborate further, a crash course in genetics, explaining the various genetic signatures and terminologies, is in order.

GENETICS 101

Changes in our DNA come about in two ways. The first takes place during DNA replication (DNA copying). When a parent cell divides to produce the germ cells, sex cells, or gametes, DNA must replicate. For a parent sex cell to divide into daughter cells (gametes), its DNA must double before division, otherwise it will lose half of its DNA every time it divides, and it will eventually perish. DNA doubles by a process called DNA replication.

But DNA replication is flawed—either by design or through an evolutionary process, it is error prone. When DNA replicates, mistakes are made in copying the original template. Most of these mistakes get corrected by the cell machinery itself, with the help of several enzymes called DNA repair enzymes. Luckily for us, only very few mistakes escape repair and get transmitted from one generation to another through sex cells or gametes (eggs and sperm) during mating. Mistakes that remain uncorrected and get transmitted from a parent to a child are about one hundred of the three billion bases that make up our genome. Our genome is a diploid genome, meaning that it is formed of two almost identical parts, each with more than three billion bases. Fortunately, most of these mistakes are not harmful.

When mutations are found at levels of at least 1 percent of the population, we call them polymorphisms. These polymorphisms differentiate humans and human groups from each other genetically and sometimes phenotypically. Their occurrence is almost exclusively generation dependent. With each new generation, novel changes are introduced. Older populations that have existed for a long time and that have produced many generations should in theory have more variations than younger populations with fewer generations. With more generations, more mating occurs, and more mating means more mutations. This is why African populations have been determined to be the oldest populations on the planet. Their genomes are the most diverse (varied) of all the populations. That is, they have sustained the largest number of polymorphisms.

The second way DNA changes is by mixing, scientifically referred to as DNA recombination. When a couple reproduce, their offspring's DNA is the result of a random shuffling of their own DNA. DNA is shuffled between chromosome pairs in the parent cell before it divides to produce the gametes.

Each of our cells contains twenty-two pairs of chromosomes and either a pair of X chromosomes (in the case of a woman) or one X chromosome and one Y chromosome (in the case of a man). This shuffling process always happens and is totally random, resulting in the production of gametes that do not resemble one another and certainly do not resemble their parent cell. In recombination, the same principle that we saw in replication mistakes holds: The older the population, the more shuffling has occurred. With every generation, there is more shuffling of the DNA.

When parents transmit DNA to their offspring, most of the Y chromosome passes as it is, without any changes from father to son. All sons will have almost exactly the same Y chromosome sequence as their father. That's because the Y chromosome does not have a homologue to pair with, unlike the other twenty-two pairs of chromosomes. All other chromosomes come in pairs, including the X chromosome. Since the Y chromosome has no homologue to pair with during male gametogenesis (when sperm cells are dividing), no genetic shuffling (recombination) can occur, as the Y chromosome has no other chromosome to shuffle DNA with. All other chromosomes shuffle their DNA during gametogenesis between homologues. Therefore the only changes or mutations that take place in the Y chromosome are those that are introduced by replication mistakes that occur when the DNA is being copied to make the gametes.

Think of it this way. Imagine that the DNA of "Scientific Adam" (since we are speaking of the Y chromosome) was written down in a letter. This letter is copied, and each copy represents a new generation. But the copying is done by typing, and typing mistakes will occur with every generation. After a few copies are made, first of the original letter, then of the copies, we end up with multiple versions of the letter. Every time we copy these versions (mating occurs, "making copies"), we introduce more mistakes. In the end, we wind up with vast combinations or alternatives of the original letter.

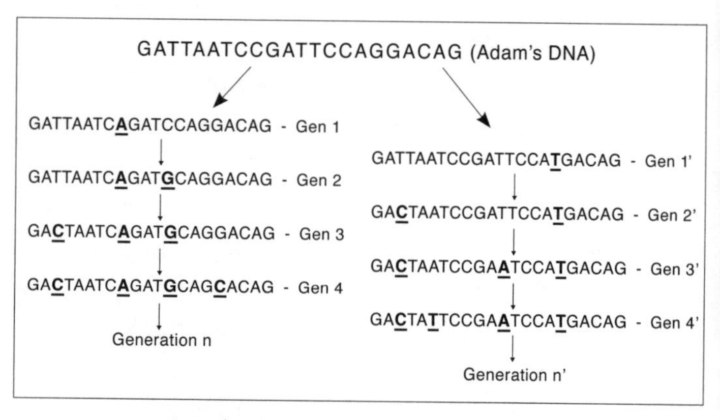

DNA replication with errors. A new mutation may be introduced with each generation.

Now the challenge will be to reconstruct backward the original letter, which we do not have (Scientific Adam is long gone), from the various circulating versions by predicting the sequential order of the mistakes that were introduced every time a version was copied. Each permutation or version of the letter is referred to as a Y haplogroup. Today more than fifty Y haplogroups and hundreds of branches (subclades) have been identified.

This is exactly how geneticists reconstruct phylogenetic trees of human populations based on Y chromosome analysis. They start with the genetic landscape of today's populations and work backward. Today our predictions have gotten more accurate, so that we can in some cases obtain the "original letter" through ancient DNA. We follow the same logic for mitochondrial DNA, which is maternally inherited.

Finally, we can also learn from genomic recombination of nuclear DNA, the DNA that is exchanged between chromosome pairs, when large sections from one version are cut and pasted onto other versions. It gets a little more complicated when we look at the entire human genome rather than those bits that are inherited via only one parent, and that is why a lot of what we know today about human migrations was initially based on mitochondrial DNA and Y chromosome studies rather than on the entire genome. Today we use all three—nuclear DNA, Y chromosome DNA, and mitochondrial DNA—in reconstructing population phylogenies.

Recently, ancient DNA analysis has proved to be an even more powerful tool. Now we can verify the reconstructions we previously made based on analyses of modern populations by comparing them to actual signatures of ancient DNA samples.

WE CAME OUT OF AFRICA IN SMALL NUMBERS, AND WE MIXED

While modern humans likely migrated in and out of Africa in many back-and-forth waves, only a couple of these waves were successful, and they were behind the spread of modern humans across the globe. Ample genetic and archaeological evidence supports the theory that modern humans left Africa through two migratory events that started somewhere in central or sub-Saharan Africa.

Y chromosome haplogroup tree showing main branches and subbranches. Terminal letters indicate the haplogroups. Haplogroup-defining mutations are indicated along the branches.

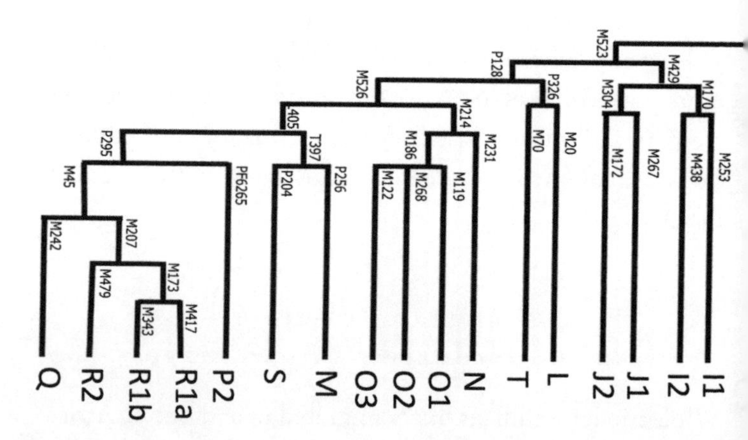

In 1987 the late Allan Wilson, a professor of genetics at the University of California, Berkeley, and colleagues published a relatively short paper in the journal *Nature* that provided the first strong piece of evidence from DNA work that our direct

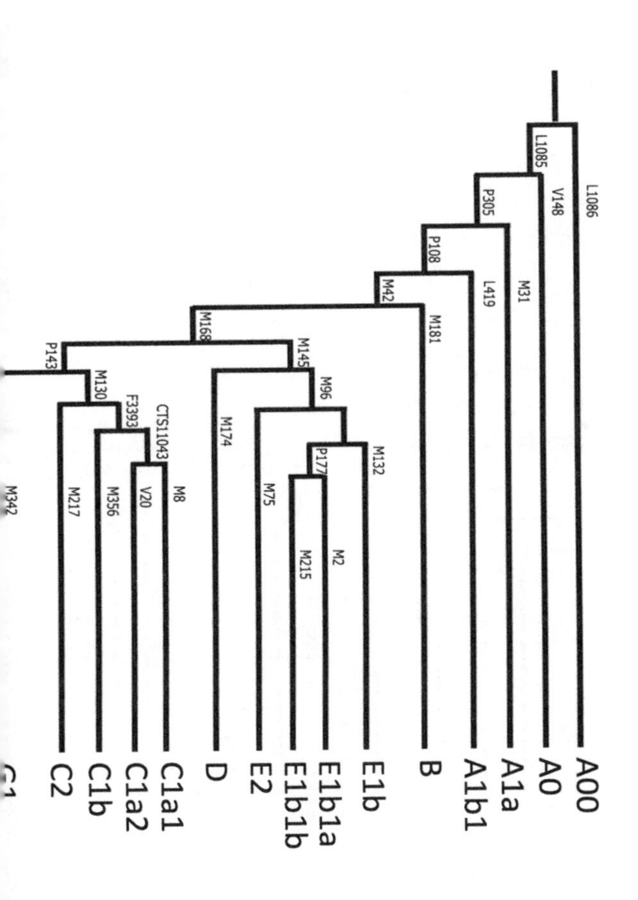

ancestors came from Africa. Their focus was on the maternally inherited mitochondrial DNA. The mitochondria are located outside the nucleus of the cell and contain their own DNA, which is passed on from mother to child without any DNA input from the father. Mitochondrial DNA basically tracks the maternal lineages.

The team analyzed the mitochondrial DNA of 147 people from various groups (Africans, Asians, Australians, Caucasians, and New Guineans) and concluded that all their maternal lineages could be traced back to a common maternal ancestor, "Scientific Eve," who lived, most likely in Africa, around 200,000 years ago. They constructed a phylogenetic tree that showed one main branch (a) yielding only African lineages and another branch (b) yielding African and non-African lineages. This b branch holds at its base the ancestor of all non-African lineages. This seminal publi-

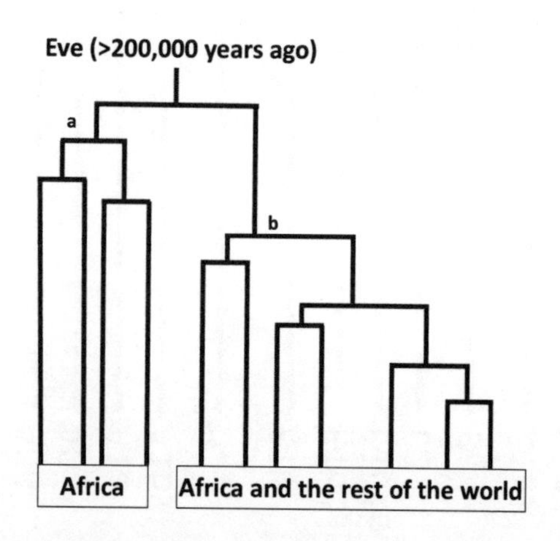

Mitochondrial DNA simplified phylogenetic tree

cation cemented the theory that all modern humans came out of Africa.

Wilson's mitochondrial DNA (maternal lineages) findings were soon supported by results from Y chromosome DNA (paternal lineages). As with mitochondrial non-African ancestral b branch, Y chromosomal DNA studies have demonstrated that modern humans left Africa more than 50,000 years ago and that all the men carried in their DNA on the Y chromosome a mutation (M168) that today is referred to as the "Out of Africa" mutation. All non-African men carry this mutation.

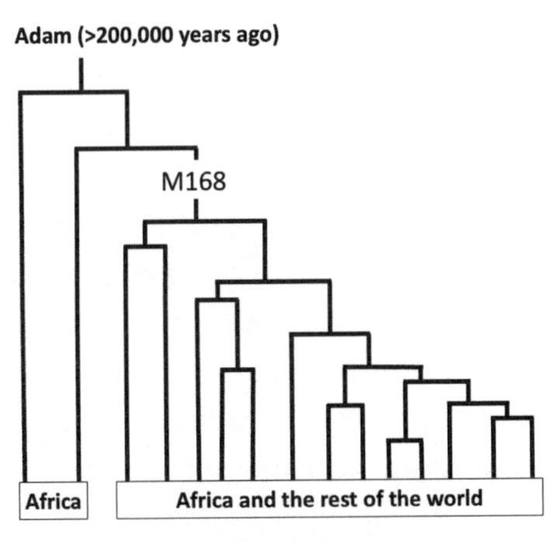

Y chromosome simplified phylogenetic tree

The overwhelming presence of this mutation in all males outside Africa suggests that at the origin of the exodus from Africa, these men experienced a serious bottleneck where only a small number of them, all carrying the defining "Out of Africa" mutation, were able to leave. If the flow out of

Africa had not experienced this bottleneck, both the mitochondrial trees and Y chromosome trees would look drastically different. The trees would have been much wider, with many more branches and subbranches and many more shorter branches across the entire trees representing early and uninterrupted admixtures.

THE SUCCESSFUL MIGRATORY WAVES

The most successful exit out of Africa may have occurred in two phases, or waves. The first may have occurred around 60,000 years ago and the second between 50,000 and 45,000 years ago. The migratory journeys these people undertook from Africa led them to the Arabian Peninsula, the Levant, Asia, Australia, and Europe, but the paths they took are not yet fully elucidated or agreed upon. Some have proposed that during the first wave, 60,000 years ago, modern humans left Africa through the Gate of Tears (Bab El Mandeb), a land passage connecting the African continent (Djibouti) to Asia (Yemen). Often referred to as the "Southern Route," it most likely brought modern humans all the way to Southeast Asia and Australia, as these early travelers followed a southeastern path along the coasts of the Indian Ocean. Milder weather conditions and suitable coastal geography may have been behind the astonishing speed and success of this spread.

The second migratory phase, around 10,000 years later, is believed to have brought *Homo sapiens* across the Sinai Desert (which was not a desert then) into the Levant.

The recent discoveries of *Homo sapiens* at Jebel Irhoud,

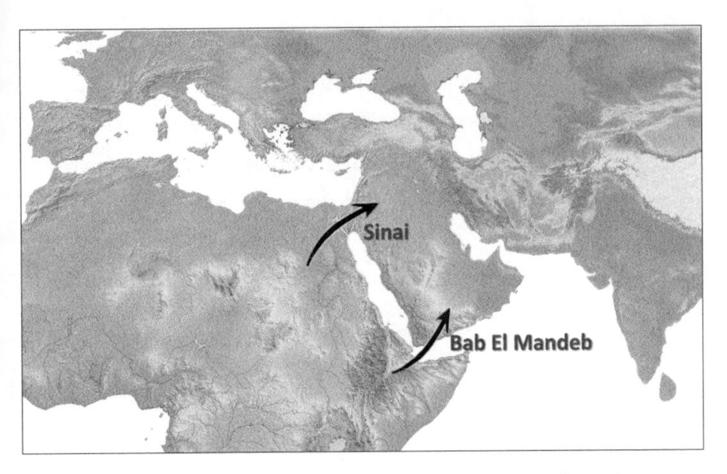

Proposed dispersion routes, based on DNA evidence by anatomically modern humans as they expanded out of Africa

Apidima Cave, and a few other locations outside Africa are now calling into question the dates of the two-wave migration theory.

Judging from the more recent archaeological record and the numerous genetic studies, the Out of Africa migration seems to have been a much more complex event than initially presumed. While these two waves may have been the most successful and important in terms of the number of individuals moving, they certainly were not the only migratory waves out of Africa. Long before our species emerged from Africa, and well before 300,000 years ago, multiple *Homo sapiens* populations were already inhabiting the entire African continent, including North Africa. That some of them might have crossed over into Asia and even into Europe should not be surprising. For the most part, however, these populations lived in isolation from one another, as demonstrated by archaeological findings, and they were morphologically diverse, although over time they occasionally interbred. From these polycentric interactions, our species eventually evolved with the set of morphological characteristics that define it and that are familiar to us today.

CHAPTER 5

Homo sapiens Meet Neanderthals in the Levant

We were never a pure breed or a homogeneous lineage

WE WERE ALREADY ADMIXED BEFORE WE "MODERNIZED"

While modern humans were taking their early steps outside Africa, a human sister group, the Neanderthals, were already thriving. Neanderthals are believed to have evolved in Africa between 600,000 and 500,000 years ago and migrated to Eurasia soon after. The first Neanderthal presence outside Africa has been documented in northern Spain, in the Sima de los Huesos cave, around 430,000 years ago. Several other locations across Eurasia document Neanderthal presence between 200,000 and 30,000 years ago.

Homo neanderthalensis and *Homo sapiens* diverged sometime in the Middle Pleistocene Period (780,000 to 126,000 years ago). They are believed to have shared, in Africa, a common human ancestor, initially thought to be *Homo heidelbergensis/rhodesiensis*. Genetic data based exclusively on mitochondrial DNA placed the species split around 400,000 years ago, a date that supports the model that considers *H. heidelbergensis/rhodesiensis* as the last common ancestor.

But new findings dispute this theory for two reasons: (1) The current autosomal DNA data suggest the split occurred between 765,000 and 550,000 years ago, or at least 200,000 years earlier than initially thought; and (2) the new dating proposed for *Homo rhodesiensis* is 300,000 years ago, as opposed to the initial date of 500,000 years ago. The search for the common human ancestor of *Homo sapiens and Homo neanderthalensis* is still ongoing.

This new date brought forth two other models to explain human taxonomy. The first suggests an earlier common ancestor, "Ancestor X," with some *H. heidelbergensis* features that existed prior to *H. heidelbergensis*. The second stays with *H. heidelbergensis* as the last common ancestor but restricts it to the oldest specimens of *H. heidelbergensis* discovered to date.

Neanderthals were shorter than *Homo sapiens,* had a larger cranial capacity, and were better adapted to cooler climates. While they appeared to be behaviorally different from modern humans, that there was some sort of genetic mixing between the two groups is now widely accepted. This mixing happened at least four times and in different locations. The first admixture event most likely took place in the Levant.

THE EVIDENCE

Up until the early twenty-first century, scientists believed that *Homo sapiens* never mixed with Neanderthals. Students in genetics courses, including mine, were taught that *Homo sapiens* and Neanderthals never interbred. We presented students with diagrams showing mitochondrial DNA recovered from Neanderthal remains that was distinctly different from mitochondrial DNA of modern humans. This was back when the only DNA available from Neanderthals was mitochondrial DNA. So these early results depicted maternal ancestry only and hence showed a clear distinction between the two groups, arguing, backed by scientific data, that the two species could not have mixed, because the mitochondrial genetic distance between them was quite substantial.

We further taught students that the two groups competed for food during harsh climate periods when resources were scarce. That led modern humans to cause the extinction of the less competitive group, the Neanderthals. More recently, new genetic findings based on full genomic data—not only mitochondrial—brought forward by a pioneer in the field of ancient DNA, Svante Pääbo, reversed most scientists' views on this matter. The more comprehensive data now support the idea that the two groups did mix and did so more than once. Not only that, but most scientists now argue that all modern humans living outside sub-Saharan Africa carry at least 1 percent to about 4 percent remnants of Neanderthal DNA in their own. Svante Pääbo won the Nobel Prize in 2022 for his work on this subject.

The percentage of Neanderthal DNA in modern humans seems to be geography- and population-specific. East Asians, for example, have the highest percentage of Neanderthal DNA, estimated at more than 2.5 percent. Europeans have less than 2 percent, most Levantine populations have around 1 percent, and Africans have nearly zero percent.

The finding that East Asians have the highest Neanderthal percentage was difficult to explain given that Neanderthals were thought to be living only in Europe at the time when *Homo sapiens* arrived there. But we now know that Neanderthals reached parts of East Asia and so were not restricted to Europe after all.

The very presence of Neanderthal DNA in modern humans has puzzled scientists, and efforts are under way to explain how, why, and when these two groups mixed. After all, Neanderthals and *Homo sapiens* have enough physical differences to have been classified as two different species. While some scientists still believe that the two are different species, some now state that the strict biological definition of species does not properly apply to closely related species, like *Homo sapiens* and *Homo neanderthalensis,* and that genetic exchange between them is not impossible.

What is fascinating is the fact that almost all modern humans today are the result of this mix. It means the offspring (hybrids) of matings between the two groups were fertile and may even have had advantages over nonhybrids. Think of the well-known example of the horse and the donkey, two distinct species: They do mate sometimes, but their offspring, the mule, is not usually fertile. Another example would be the product of the mating between grizzly and polar bears, the "grolars." Grolars, unlike mules, are usually fertile, but grizzly and polar bears are considered subspecies. In any case, today the scientific evidence shows clearly and abundantly that *Homo sapiens* and Neanderthals did interbreed, and that they did so at least four times, in different geographical locations, and at least over a 20,000-year period.

THE MIXING

While *Homo sapiens* may have briefly interbred with Neanderthals at some earlier point during their many "unsuccessful" outings from Africa, one widely accepted theory is that soon after their successful exit from Africa 60,000 years ago, modern humans seeking refuge from the harsh weather conditions encountered Neanderthals occupying some of the same refuges in the Levant and interbred with them.

The Levant is almost void of any archaeological record of *Homo sapiens* between 80,000 and 60,000 years ago, but numerous sites throughout the Levant show continuous evidence of Neanderthal fossils up to 45,000 years ago. When around 60,000 years ago *Homo sapiens* arrived in the Levant, as the weather shifted to favorable conditions, the Neanderthals interbred with them. This encounter happened early on during this migration wave, and given the small number of *Homo sapiens* taking part in this migration, all their descendants now carry Neanderthal DNA. This also explains why sub-Saharan Africans have little or no Neanderthal DNA.

Proponents of this theory further claim that some of the Neanderthal DNA acquired by *Homo sapiens* provided additional protection against infections; hence only *Homo sapiens* with mixed Neanderthal DNA were able to survive and thrive. According to recent findings, some of the Neanderthal genetic variants that got fixed in *Homo sapiens* genomes are indeed beneficial and appear to confer better immunity against certain viral infections, like Covid-19. Additional explanations have also been proposed to explain the varying

Neanderthal percentages among modern humans. Some researchers believe that, as these "hybrids" moved into Europe, they further interbred with an unknown local population (most likely *Homo sapiens* from earlier migrations) that diluted their Neanderthal percentage. That admixture did not happen in East Asia, explaining the higher percentages there. The same dilution theory has been proposed to explain the low Neanderthal percentages in Levantine populations. Other scientists attribute the varying percentages to multiple admixture events in East Asia, or they propose that selection favored Neanderthal genes differently in different environments.

What remains peculiar about the interbreeding theory is the absence of evidence showing that *Homo sapiens* DNA made it into Neanderthal genomes, suggesting a one-way DNA transfer, from Neanderthal to *Homo sapiens*. Several explanations could be put forth: (1) Not enough Neanderthal genomes have been sequenced to prove conclusively the absence of *Homo sapiens* DNA; (2) interbreeding was more frequent within *Homo sapiens* communities than within Neanderthal communities; and (3) *Homo sapiens*–Neanderthal hybrids within the Neanderthal communities were rare or even nonviable. Given the absence of Neanderthal mitochondrial DNA from *Homo sapiens'*, most of the interbreeding could perhaps have been between Neanderthal men and *Homo sapiens* women. Or perhaps only the descendants of matings between Neanderthal men and *Homo sapiens* women were fertile, and not the descendants of the reverse pairings. If so, this would explain why more fertile Neanderthal men were assimilated into communities of *Homo sapiens* than the reverse, which may have partly led to the Neanderthals' eclipse.

THE DEMISE OF THE NEANDERTHALS

Several factors may have favored *Homo sapiens* over the Neanderthals, though the exact reason for the latter's extinction remains to be fully elucidated. Did Neanderthals become extinct because of their competition with *Homo sapiens* for resources, because they lacked some genetic adaptation to extreme changes in climate, or for some other reason? It remains unclear. Recent evidence, however, shows that *Homo sapiens* and Neanderthals coexisted for longer periods than was initially thought, and they may have shared the same geographical locations in Eurasia before the Neanderthal demise 30,000 to 25,000 years ago.

Several theories about what led to the Neanderthals' extinction have been proposed. The climate was constantly shifting between cold snaps and very hot temperatures, and these shifts were extreme, without a transitioning phase for human adaptation. The weather alternated from wet to dry within short time periods, leaving the Neanderthals with fewer hunting grounds and no time to significantly change their hunting habits or adapt to the abrupt climate variations. The warmer climate also led to the decline of the Neanderthals' main hunted prey, the large woolly mammoths and cave bears. *Homo sapiens,* on the contrary, while in Africa evolved their hunting skills and developed more complex hunting methodologies and versatile tools. They were better able to cope and survive in variable climates. In the end, the

Neanderthal social structure was no match for that of *Homo sapiens,* who competed for almost the same resources. Most likely the combination of climate shifts and their more developed hunting skills favored *Homo sapiens* to thrive at the Neanderthals' expense.

While the interbreeding theory is strongly supported by DNA analysis, some scientists still claim that the two groups did not interbreed at all. Neanderthal DNA is present in *Homo sapiens,* they argue, because the two may have had a very recent common African ancestor that is yet to be discovered. The latter theory is supported by archaeological evidence based on new methods of dating showing that Neanderthals were only present in Eurasia before modern humans arrived, and during that period the two groups did not cross paths.

DENISOVANS AND *HOMO SAPIENS*

Neanderthals were not the only human group that shared the globe with *Homo sapiens*. Another group, the Denisovans, existed some 300,000 years ago in East Asia. Little is known about them, but the Denisovans, like the Neanderthals, may have shared a common human ancestor with *Homo sapiens,* whether it is *Homo heidelbergensis* or Ancestor X. It is believed that the Denisovans split from the Neanderthals soon after (within 100,000 years) the latter split from *Homo sapiens*. While the Neanderthals occupied Eurasia, the Denisovans were geographically restricted to Asia. They coexisted and interbred with *Homo sapiens* in East Asia at least twice, as is attested by the presence of Denisovan

DNA in some modern East Asians. The highest levels of Denisovan DNA (as high as 5 percent) are found in Australian Aboriginal and Melanesian populations. Modern humans may have encountered the Denisovans in Siberia and South Asia at two or three separate events. In other words, modern humans mated with two different Denisovan populations and in two geographically different locations.

The scientific debate is ongoing, and it will be awhile before we really know exactly how modern humans integrated Neanderthal or Denisovan DNA into their own. We are genetically part Neanderthal, part Denisovan, and part *Homo sapiens,* but we are all 100 percent human.

PART III

THE EARLY SETTLEMENTS

CHAPTER 6

Our Early Ancestors in the Levant

Human nature is not of the settling kind

THE GENETIC EVIDENCE

Soon after modern humans' arrival in the Levant, they began to establish small groups that developed into small hunting communities. But the Last Glacial Period was marked by extreme climate fluctuations, between significant cooling and aridification, which throttled modern humans' expansion and forced dispersed human populations into geographically confined shelters or refuges with more suitable climates.

Archaeological evidence supporting this theory of intermittent dispersal during the Last Glacial Period is confined to a handful of geographically limited human population sites in the Levant. Some of these shelters or refuges show archaeological evidence of distinctive tool cultures: in the Levant, in the Mediterranean coastal region of the Taurus Mountains (in eastern Anatolia), and in the Zagros Mountains (along the Persian Gulf coast).

These refuges provided modern humans with good shelter from the extreme cold that prevailed during that period and allowed some vegetation to grow that could support wildlife. But throughout the cold spells, they could not easily mix with one another, as their movements were severely constrained. Due to the harsh weather conditions, these early *Homo sapiens* hunting communities remained relatively isolated from each other.

Unique genetic signatures, including specific mitochondrial DNA and Y chromosome signatures, would have evolved in these relatively isolated communities by a phenomenon called genetic drift. Genetic drift occurs when specific DNA signatures (including a set of genetic traits) characterize and distinguish population isolates. These traits become even more frequent in the isolated population due to inbreeding.

That the Levant and other parts of Southwest Asia constituted a permanent habitat for all these early migrants is highly unlikely. During the peak of the Last Glacial Period, most of the Levant and the Arabian Peninsula were arid and uninhabitable. The water levels of the Persian Gulf basin were at their lowest, about one hundred meters below current levels, which left almost the entire peninsula dry, forcing the early indigenous groups to seek refuge elsewhere. While some of these early populations retracted into refuges across the mostly exposed basin, others likely ventured farther north (into Mesopotamia/Iran) or northeast (Anatolia), or they back-migrated into Africa (Ethiopia).

Around 18,000 years ago, the ice started to melt, and the waters of the Persian Gulf basin rose, flooding the surrounding plains and reaching their peak extent around 7,000 years ago. Meanwhile the entire region from the Levant to Arabia witnessed a population boom. With the melting of the ice caps, temperatures also rose, and the semi-sedentary hunter-gatherers in the Caucasus, the Zagros Mountains, eastern Anatolia, and the Levant dispersed from their localized habitats along the coast of the newly reshaped Persian Gulf and into Arabia. This expansion scenario is consistent with the archaeological evidence demonstrating a sizable increase in the populations of southern Arabia around 7,500 years ago.

Postglacial warming (and later the demographic consequences of agriculture) favored the growth of these populations, with outcomes that are visible in present-day patterns of genetic variation among peoples. The first Neolithic farming communities most likely followed the specific routes of expansion used by the early inhabitants of the Levant.

Archaeological evidence shows that post–Last Glacial Period expansions and their aftermath were marked by increased social complexity, a shift from mobile hunting and gathering to more sedentary lifeways, the emergence of trade and exchange networks, and the early cultivation of wild cereals.

THE GATE OF TEARS PATH AND THE MAKEUP OF ARABIA

Soon after the Last Glacial Maximum, 23,000 to 18,000 years ago, *Homo sapiens* settlers (after earlier back-migration events) migrated from East Africa through the Gate of Tears into Yemen. From Yemen, some moved into the rest of the Arabian Peninsula, while others followed the coast along the Persian Gulf, reaching Iran. The water levels of the Persian Gulf back then were at least twenty meters lower than they are today.

While the Levant's optimal climate was supporting population growth, Arabia continued to witness severe weather fluctuations, including flooding. Recent archaeological findings from Wadi Hilo in southeastern Arabia provide evidence for human occupation since the Early Holocene, around 11,500 years ago. The early occupants of Arabia who appeared

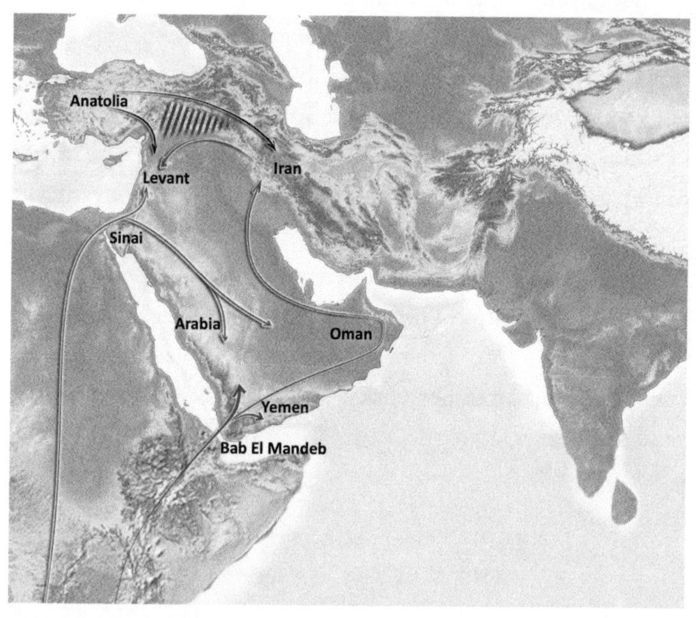

Migration paths of Southwest Asian ancestral populations. Arabian Peninsula to Persian Gulf and Iran lineages' path through the Gate of Tears; Natufian and Levantine lineages' path through the Sinai; pre-Neolithic Anatolians' path into Iran and the Levant; Neolithic Iranians' path into the Levant. The Neolithic Core Zone (see chapter 7) is indicated by the striated area on the map.

after the Last Glacial Period had variable lithic tools, as well as locally specific hunting habits, and made rock art in the Hejaz. Many more people from East Africa, the Levant, or Mesopotamia arrived in Arabia. Some of these tool cultures indicate Levantine dispersal, while others suggest autochthonous development, and still others favor a Levant-independent African arrival.

The first inhabitants of Arabia from 60,000 years ago, contracted into the shoreline refugia of Arabia, had to face the water expansion and flooding of the Persian Gulf basin. Whether they survived remains to be investigated. If some of the early migrants did survive, then Yemen and the rest of southeastern Arabia undoubtedly witnessed additional population influx from the various post–Last Glacial Period expansions that were occurring throughout the Levant, Anatolia, Mesopotamia, and perhaps East Africa.

The modern populations of the Arabian Peninsula carry genetic signatures marking mainly the pre–Last Glacial Period migration from Africa to Arabia and Iran. DNA analysis argues also for a post–Last Glacial Period flow of Levantine lineages south through Arabian Peninsula populations toward Yemen and not the reverse. Overall, these analyses are consistent with early southern coastal settlement across the Arabian Peninsula toward the Persian Gulf and Iran and a northern settlement from the Levant and Anatolia. Neolithic Iranian lineages are present in Yemen and Qatar but are not as significant in other populations from the Arabian Peninsula, indicating a primarily coastal settlement reaching Iran. The absence of the genetic flow of Iranian lineages from Yemen to other Arabian Peninsula populations argues for homogenization of these lineages within the Arabian Peninsula over time.

The Omanis carry ancient genetic signatures distinct from all other ancient populations of the region. In addition, the Omani population appears to have a component linked to Africa in what may represent the genetic signature of historic interaction and trade networks. The Luhya, a Bantu-speaking population sampled in Kenya, are one likely source population for some of the Omanis. This pattern is consistent with Oman's historic involvement in the Bantu slave trade that focused on Southeast Africa beginning in the seventh century CE.

The ancestral populations that gave rise to the modern populations of the Arabian Peninsula, however, remain largely uncharacterized. All this points to multiple complex population movements from various post–Last Glacial Period refuges that led to the makeup of the current Levant, Mesopotamia, and Arabia. These movements likely involved ancestral populations that have long since disappeared, leaving no detectable genetic trace, while others' genetic contribution is still sustained in modern populations.

While the ancient DNA remains of early migrants into Arabia have yet to be discovered, they are believed to be the ancestors of the modern inhabitants of the Arabian Peninsula. Some Arabian populations harbor unique genetic signatures that are yet to be detected in other populations. Perhaps the bearers of these unique genetic signatures succumbed to the Persian Gulf waters following sea level rise after the Last Glacial Period. These may, one day, be discovered under the sea.

THE SINAI PATH AND THE MAKEUP OF THE LEVANT

Another group of *Homo sapiens* left Africa at a date later than the one proposed for those who populated Arabia and the Persian Gulf. This group took the Sinai path into the southern Levant. From the southern Levant, a group of them continued their journey to the northern Levant, where they interbred with the Neolithic Iranians coming from the Zagros Mountains refuges.

THE ARCHAEOLOGICAL EVIDENCE

Around 18,000 years ago, three early foraging cultures appeared in several small geographical pockets: the Kebaran foragers in the Levant, the Belbasians in the Anatolian Steppe, and the Zarzians in the Zagros Mountains. As the climate warmed, these early foragers expanded to sites farther south, where vegetation was lusher and wild game more abundant. Around 14,000 years ago, these small, stable groups abandoned foraging and developed a semi-sedentary lifestyle; then they left behind the semi-sedentary lifestyle in favor of permanent settlements. These early settlements were the hallmark of the cultures that appeared soon after 14,000 years ago in the Levant (the early Natufians), in Anatolia (the Beldibians), and at the foot of the Zagros Mountains (the Zawi Chemi). These are the first cultures known to have adopted a sedentary lifestyle since *Homo sapiens* left Africa.

While the Natufians gave rise to the early farmers in the Levant, neighboring populations that occupied the northern Anatolian and the eastern Iranian plains gave rise to the early Anatolian and Iranian farmers who later expanded from their refuges into Eurasia.

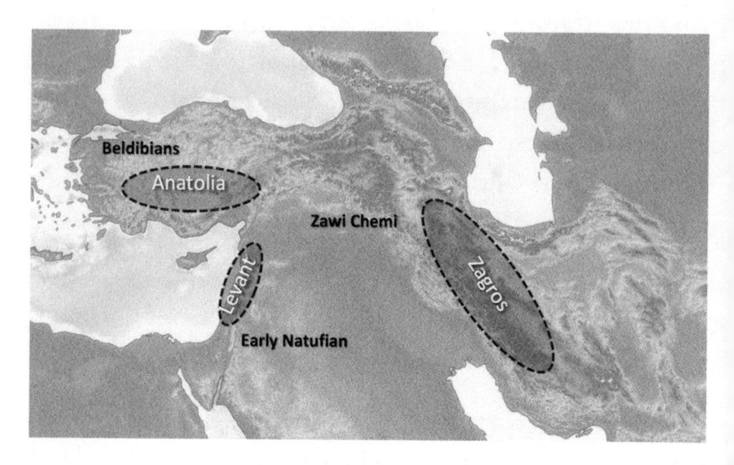

The early cultures and their locations

THE NEOLITHIC TRIANGLE

DNA analyses performed on ancient human remains collected from Southwest Asia show that the Natufians, the Beldibians in Anatolia, and the Zawi Chemi in the Zagros bear distinguishable genetic patterns. The population centers that gave rise to the Natufians, Neolithic Anatolians, and Zawi Chemi were isolated from one another, evolving separately in their geographical locations long enough to develop distinct genetic features before they eventually mixed. These early populations constitute the "Neolithic Triangle," linking the Levant to Anatolia in the west and the Zagros Mountains to the east.

THE NATUFIANS

Given its geographical location, the Levant constituted an important transition zone for the early hunter-gatherer migrants. By the time *Homo sapiens* developed from hunter-gatherers into largely non-nomadic farmers, genetically distinct population centers, or isolates, already existed throughout Southwest Asia, including the Levant and Arabia.

The Natufians may have survived the last ice age by sheltering in the small refuges that existed throughout the Levant; after the ice sheet melted, they spread out from these refuges. The warming climate in the Levant brought more precipitation and with it an abundance of vegetation and wild animals. This abundance of food attracted Levantine foragers to occupy more southern sites. Their early settlements consisted of small dwellings around Wadi Al Natuf (hence the name Natufian), located in Ramallah about thirty kilometers northeast of Jerusalem. The early Natufians settled mostly around woodlands in the southern Levant, where open forests were also available and cereals grew in abundance. Their early habitat areas did not exceed one thousand square meters. Settlements have also been discovered in the northern Levant, in the Bekaa Valley, and in Beirut, albeit fewer in number.

The Natufian populations in the Levant experienced a relatively pleasant environment for about two thousand years, although it was not void of periods of harsh, abrupt, and extreme weather shifts that may have played to the Natufians' adaptive advantage. As they adapted further to climate change and moved from areas of less food abundance to areas of more abundance, they acquired more adaptive skills. They moved as groups and never went back to the foraging hunter-gatherer lifestyle. These abrupt micro-climatic shifts of limited duration prepared the Natufians well for a major dip in temperature between 12,900 and 11,600 years ago, which left the southern Levant and Arabia almost entirely arid. This period is called the Younger Dryas, in reference to the *Dryas octopetala,* a plant that grows in extremely cold conditions and emerged in Europe during this period.

The climate shift of the Younger Dryas led to increased aridity and food depletion in the Natufians' adopted homeland in the southern Levant. The area became uninhabitable, leading to a major dip in the Natufian population size. The Natufians (now referred to as Late Natufians) were forced to move from the southern Levant in search of wild cereal pastures and new hunting grounds and to improve their hunting skills. They found new shelters in the Negev highlands, the northeastern Levant, and Mesopotamia. In the Negev highlands, they experienced further adaptation to aridity and became referred to as the Harifians.

THE EARLY HABITATS

The calamitous weather conditions of the Younger Dryas in most of Southwest Asia left the Natufians and other sedentary and semi-sedentary dwellers with little food. The dry conditions were a catalyst for these peoples to develop new skills in order to cope.

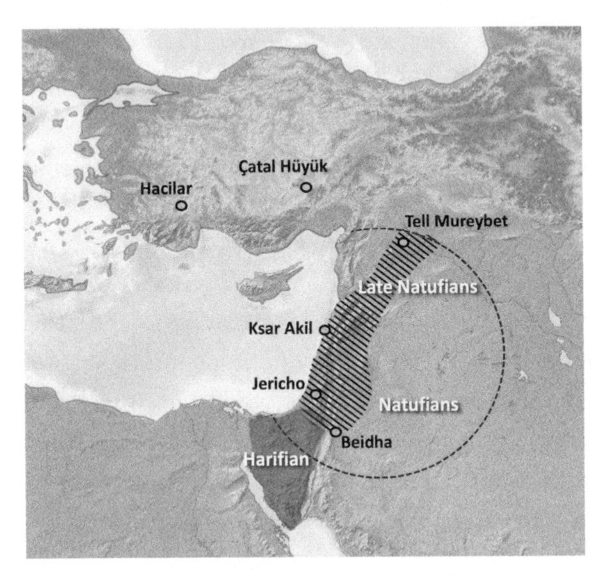

The Natufians and their early habitats

During this period, humans experimented with planting and migration, seeking better conditions for their small crops of mainly cereals—barley and wheat, depending on their location. Barley grows in temperate conditions, while wheat can sustain cooler weather. As the climate conditions improved further, wild cereals and vegetation flourished, drawing humans out of their refuges. They used larger areas for cultivation, and around 10,000 years ago, the first villages appeared around these cultivated areas, with a few hundred inhabitants each.

These villagers developed community life around places of worship, domestic rituals, and other community activities. From this point onward, they favored farming and domestication over hunting. As the process of domestication expanded their living areas, they moved across the Levant to establish more farming communities. The Beidha site in Jordan, near Petra, was occupied from the Mesolithic Period, or Middle Stone Age (10,000 years ago), and continued to be occupied uninterrupted for nearly 7,000 years. Beidha is situated on high ground, a shelf of land that is about one thousand meters above sea level overlooking the Jordan Valley; the site shows evidence of cultivation of barley and emmer wheat but no pottery. The people who occupied this site used containers made of wood and stone for storing cereals and other grains. Beidha was not isolated but was one of multiple similar sites along the western arc of the Levant, the most famous of which was Jericho.

CHAPTER 7

From Anatolia and the Zagros to the Levant

Agriculture, a double-edged sword with a poisonous blade

THE NEOLITHIC CORE ZONE

The Natufian expansion area in the northeastern Levant coincided with human activities in the western portion of the Fertile Crescent, where wild cereals grew in abundance. The Natufians encountered and admixed (interbred) with the early Anatolians (Beldibians) who also came out of their refuges after the last ice age. The Beldibian culture emerged in Çatal Hüyük and Hacilar, in southern Turkey, around the same time as the Natufians. Together these two groups constituted most of the population of the Levant during the Neolithic Period, or the New Stone Age Period (12,000 to 4,000 years ago).

Neolithic Anatolia contributed substantially to Levantine populations. DNA evidence shows that the Neolithic Levantine populations derived almost two-thirds of their genetic ancestry from the Natufian lineages and the remaining one-third from the Neolithic Anatolians and, to a much lesser extent, the Neolithic Iranians.

The Neolithic Iranians were early farmers, some of whom

migrated from the Zagros Mountains to the Levant during the Neolithic expansion. Their early ancestors may have been the earliest settlers of Arabia and the Persian Gulf, who came from Africa through the Gate of Tears route, likely following the coast, and ultimately reaching the Zagros Mountains in the north (discussed in the previous two chapters).

Neolithic Iranian lineages, which are also present in the Arabian populations, spread into the Levant well after the Natufians were firmly established there. Their movement to the Levant occurred much later than the Neolithic Anatolians, and happened in much smaller numbers.

While the Anatolian spread to the Levant was accompanied by increased genetic flow, the Iranian admixture into the Levant likely occurred directly from Iranian settlements at the foot of the Zagros Mountains. Neolithic Iranians expanded into the Neolithic Core Zone, a triangular area stretching from the Levant in the south to eastern Anatolia and through northern Mesopotamia into the Zagros Mountains. To get to the Levant, they followed a northern path. They did not go through Yemen and the Arabian Peninsula but went along the Tigris and Euphrates rivers. If the Iranians had used the Arabia route, then Arabian lineages would have been carried along the way, but these are not found in the Levantine populations.

The expansion of modern humans out of Africa into the Persian Gulf region appears to have left a genetic legacy in the Arabian Peninsula. The path from Africa into the Levant shows no similar legacy in the Peninsula, suggesting a totally different route, the Sinai path.

From the Last Glacial Period to the present day, the Levant has been an attractive region, desired for its temperate climate and its lush vegetation and wildlife. Together these three ancestral populations accounted for the main genetic stock of the Levant up until the Early Bronze Age (5,400 to 4,400 years ago). Over the last four millennia, however, many populations have moved in and out of the Levant, and while each one has left its cultural mark (art, language, and much more), their genetic footprints remained fairly limited.

THE STEPPE (YAMNAYA) PEOPLES AND THE LEVANT

Another group of people involved in the genetic makeup of the Levant were the Steppe peoples, whose history is still being mapped. Identified first in Europe, the Steppe peoples, particularly the Yamnaya culture, have been dated to around 5,000 years ago and are credited with bringing the Indo-European language to Europe (the Kurgan hypothesis). Their appearance in Southwest Asia has been associated with the Hittites, who are thought to have brought the Indo-European language into the Fertile Crescent around 3,000 years ago. The Kurgan hypothesis posits that the Pontic Steppe peoples brought Indo-European languages with them as they expanded, with some ultimately reaching Anatolia. This hypothesis is supported by genetics and deviates from previous theories that regarded Anatolia as the center from which the Indo-European languages expanded.

The Indo-European languages arrived within the historic

time frame of the Hittites and similar groups in Anatolia. Prior to that, Anatolia was home to a few non-Indo-European isolates such as the Hattis, Hurrians, and Urartians. Notably, the Steppe admixtures present in the Natufians and Neolithic Iranians predate the Hittites.

HOW DID THE YAMNAYA LINEAGES MAKE IT TO THE LEVANT?

The Yamnaya genetic lineages are derived from European hunter-gatherer genetic lineages, representing both Scandinavian and Western European hunter-gatherers. The Yamnaya lineages likely admixed with Natufians and Neolithic Iranians well before the Hittite expansions between 3,500 and 3,200 years ago. Curiously, these Yamnaya lineages are not found in the Arabian Peninsula, suggesting that while they made it to the Levant and to Iran, they did not reach the Arabian Peninsula. Even more interesting, the Anatolian lineages also did not significantly spread into the Arabian Peninsula. These observations argue that the early inhabitants of Anatolia likely carried the Yamnaya lineages into the Levant.

NEOLITHIC IRANIANS INTO ARABIA

The Neolithic Iranians carried the Yamnaya lineages (admixed with each other), but the Arabian Peninsula populations that admixed with Iranian lineages do not carry those Steppe lineages. Clearly, the Neolithic Iranians, after acquiring Yamnaya lineages, did not play a role in establishing these lineages in Yemen and other parts of the Arabian Peninsula. So when did the Neolithic Iranians (minus the Steppe lineages) mix with the Arabian Peninsula populations? If the Neolithic Iranians arrived in Arabia from the north (current Iran), their admixture with the Arabians had to have happened prior to the Neolithic Iranian admixture with the Steppe lineages, which is unlikely. More likely, the early Iranian lineages came through the Arabian Peninsula—originally from Africa (evolving on the way)—and went north, where they mixed with the Steppe lineage–carrying populations.

THE AGRICULTURAL REVOLUTION

The Agricultural Revolution was born farther east, across the Deh Luran Plain, in southwestern Iran. The Agricultural Revolution is also called the Neolithic Revolution in reference to the more refined lithic (stone) tools, like chopping axes, grinding stones, and sharper arrowheads, that the early farmers used. It marks the major shift by *Homo sapiens* from hunters and gatherers to farmers starting around 12,000 years ago. And it marks the beginning of population expansions not only in size but in territory. At sites across the Deh Luran Plain, semi-sedentary peoples with seasonal site occupation gradually gave way to farmers in permanent settlements. The area's fluctuating climate, in which severe cold snaps changed to temperate spells and back again in a short period of time, forced people to adapt to harsher environments and may have played an important role in this transformation. Multiple environmental zones can be found within this region, ranging from mountains and dry steppes to marshes and sedimentary plains. It is surrounded by rivers and supports a wide variety of wild game.

In the Zagros Mountains of northern Iraq, as temperatures were getting warmer around 10,000 years ago, cereals and other vegetation started to appear, leading to the emergence of several Neolithic cultures, like the Zawi Chemi/Shenidar, along the foothills. This culture, unlike the contemporaneous Levantine cultures, resided in seasonal sites, moving to lower altitudes during the winter months. When this culture extended from northern Iraq to the Deh Luran Plain in Khuzistan (modern southwestern Iran), it underwent a gradual transition from semi-sedentary and seasonal settlements to permanent settlements.

Neolithic sites extended farther along the Irano-Turanian Plain's vegetational belts. Tell Mureybet is one of the largest Neolithic habitats, situated about eighty-six kilometers east of Aleppo, Syria, between the Levant and Mesopotamia. This site was first inhabited around 10,500 years ago, and with time its population expanded, reaching the unprecedented size of approximately one thousand, living in about four hundred individual dwellings. Tell Mureybet was occupied without interruption for more than 2,000 years. It is considered the largest Neolithic dwelling site, with round houses ranging from about 2.5 to 6 meters in diameter. What may have favored the subsistence and survival of people there was the site's proximity to fresh water, being situated on the left bank of the Euphrates.

This site also witnessed the presence of fire and roasting pits, surrounded by several houses. Two or three small villages were first established on small circular mounds not larger than 150 meters in diameter, and from these, other villages sprang up, until the entire plain was populated sequentially. With time, reliance on farming dominated while herding and hunting subsided, marking a major shift in diet with much less dependence on meat. Houses became larger (five by ten meters), with internal partitions and hinged doors accommodating several individuals per household. With time, ceramic cups and containers appeared, as well as human figurines and mural paintings. Houses were built directly on the ground, their walls constructed with a puddled clay mixture made of earth, stone, straw, and bones. A remarkable rectangular building with four rooms was excavated at Tell Mureybet, most likely a place of worship.

THE FIRST CULTURES

Starting around 9,000 years ago, multiple tiny cultures dotted an area stretching from the Levantine Corridor and the Jordan Valley to the Damascus basin and the Anatolian Steppe to the north and the Irano-Turanian Plain to the east. This area corresponds closely to the Fertile Crescent. These cultures were the main catalysts of the Agricultural Revolution. The early farmers were none other than the Natufians, the Zawi Chemi, and mostly the Beldibians, who had been dwelling in the Fertile Crescent for several thousand years, evolving their skills and ways of life until they adopted farming. European archaeologists referred to these early farmers as Neolithic based on the stone tools and stone industry they developed.

Eight thousand years ago village life in Mesopotamia was in full swing, and for the first time in the world, at least three distinct cultures appeared: the Hassuna, the Samarran, and the Halaf. The basic element of these cultures was the nuclear family, which constituted the first social unit within a household. Many more families formed around each other, creating a social organizational structure within a confined geographical space that we now call a community from a social perspective and a village from a structural perspective. These were the defining cultures that shaped humanity and made it the way it is today.

Broadly, these cultures all evolved between 8,000 and 7,000 years ago. The Samarran culture occupied southeastern Mesopotamia, the Halaf culture occupied northwestern Mesopotamia, and the Hassuna culture was restricted to northeastern Mesopotamia, in what is now northern Iraq. The villagers grew einkorn, emmer, and club wheat as well as lentils, naked and two-row hulled barley, linseed (flax), and peas in much larger quantities and greater varieties than in any other sites before. They domesticated cows, sheep, goats, pigs, and dogs, hunted fallow deer and gazelles, and exploited the Tigris and Euphrates for fish and mussels. The Samarrans invented irrigation, in its simplest form, while the Halaf culture is considered the most agriculturally advanced of them all.

The Hassuna culture produced female figurines, fine painted pottery, and jar burials with food remains and other tools for an afterlife. With its small dwellings, it represented the first truly agricultural village-life in northern Mesopotamia. The Halaf culture was notable for its finely painted pottery and its use of *tannours,* or dome-shaped bread ovens (still widely used in today's Levant), and of *tholos,* well-developed circular-shaped houses that often replaced early rectangular structures. The Halaf culture spread and influenced the Samarran culture that later transitioned into the Ubaid culture, and both expanded farther south into Mesopotamia. Around 6,000 years ago these two cultures transitioned into the Uruk Period (c. 6,000 to 5,100 years ago), when the first well-developed cities emerged and with them the first dynasties of Sumer.

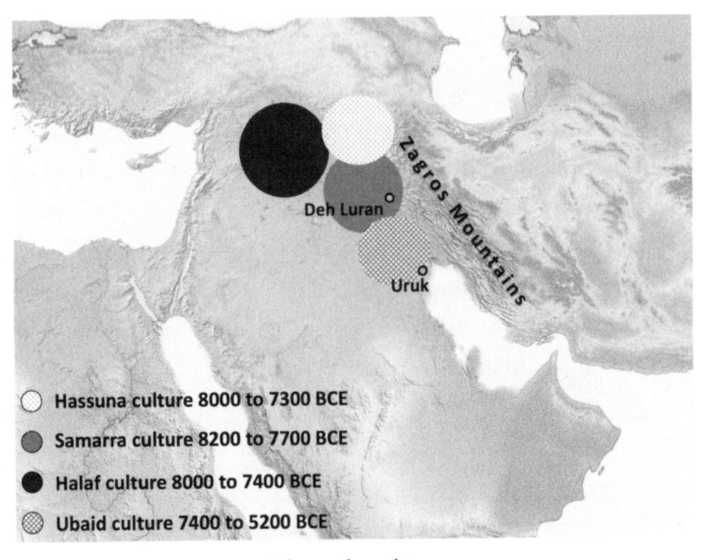

The early cultures

CHAPTER 8

The Levant in the Neolithic Period

Seasonal variation is not a luxury

As the climate continued to improve, farming replaced hunting as the main food source for Southwest Asian communities. As we have seen, around 12,000 years ago, farming communities emerged slowly in several spots in the Levant and Mesopotamia, ushering in the Neolithic Period, when more refined stone tools (hunting axes and arrowheads) were developed. These communities' habits and rituals would develop into the full-blown cultures that define the populations of today.

FARMING IN THE LEVANT

When I was growing up in the western Levant, seasonal variability seemed a natural occurrence. Spring and fall are colorful, winter is rainy and cold especially in the mountains, and summer is hot and relatively humid. Another natural occurrence was that at the end of each school year, usually in early June, we moved to our summer mountain village, located about twenty kilometers away. We would spend the entire summer there, where the weather is cooler, without the uncomfortable humidity of the coast. Our summer village is located at an altitude of 1,450 meters above sea level and is surrounded by one of the most biodiverse temperate forests in the world. This forest, with its many phytoclimatic zones, covers a small area of less than one thousand acres of land, but it is a haven for more than one thousand species of plants including about forty species of trees, the majority of which are native to the eastern Mediterranean.

Four seasons and a summer home are typical for a Mediterranean climate, defined as a temperate climate where the four seasons are observed.

In his 1997 book, *Guns, Germs, and Steel,* Jared Diamond talks about five advantages of the Fertile Crescent (the Levant included) that made it the most suitable site for early human development. One is its varied seasons, and another is its varied topographies within a small, geographically confined region. The others have to do with the region's biodiversity. In today's Levant, especially in the western part, the seasonal variability persists. We still witness the four seasons with almost perfect precision, and the areas that have escaped random urbanization and land destruction are extremely rich in biodiversity.

What is most striking, however, is the concept of a summer village. When my family moved every year to our mountain village for the summer, we did not move alone: Everybody in my hometown moved, and that included all domesticated animals (sheep and goats) and the beehives as well. In fact, the two towns belong to the same people, they constitute one municipality, and you cannot belong to one without belonging to the other. You cannot be the resident of only one—once you reside in one, you are de facto a resident in the other.

In my coastal town, mostly olives are grown, with some oranges. To maximize their crops, my direct ancestors (and perhaps their ancestors before them) grew a variety of crops, taking advantage of the region's variations in seasons and topography. It seems that this behavior has been going on for many thousands of years. In fact, the farmers of northern Luristan, in the Zagros Mountains, have a somewhat similar custom today. But they move four times a year into four different dwellings. In the winter they occupy stone houses called *zemga* in their primary village. In the spring they move to their *siah cador* (black tents), in the summer they move into *kula,* booths constructed of tree branches, and then in the fall they go back to tents. These four dwellings are only a few kilometers apart but are located at different altitudes with significant variations in temperature and environment. The search for favorable agricultural and pastoral opportunities is what drives the seasonal movement of these Luristan farmers. For the very same reasons, my ancestors and those of the people of Luristan developed similar customs.

THE EARLY CROPS

Ten thousand years ago, the Levantine inhabitants cultivated eight cereal and grain crops that were considered symbols of the Neolithic Revolution. Six of them are still abundantly cultivated today and are used in daily meals across the Levant and beyond. They still constitute the main staple foods in all of Southwest Asia. These are einkorn and emmer wheat, lentils, chickpeas, peas, and barley. The remaining two, linseed (flax) and bitter vetch, are still present but are not used as food sources, nor are they heavily cultivated.

That these grains are used daily in such a large variety of dishes is testimony to their long presence in the Levant. People around the world are now familiar with falafel, hummus, and tabbouleh. It is remarkable that these culinary traditions were passed on uninterrupted for so long, and how easy it is to just accept them as natural occurrences and not as evolutionary events that determined our continuity as a culture.

PART IV

A COMPLEX GENETIC MAKEUP

CHAPTER 9

Population Expansions

Blame it on climate change—
spread those genes

CHAPTER 8 DESCRIBED HOW AND when modern humans arrived in the Levant. This chapter will explore how they genetically evolved within their early habitats during the Last Glacial Period before they expanded, as is demonstrated by their level of genetic diversity. When a group had enough time to exchange genetic material by mating, over multiple generations, individuals in that group become genetically more diverse.

The current genetic landscape of the Levant and the rest of Southwest Asia follows patterns that are consistent with multiple populations that underwent long periods of isolation before they expanded. DNA work on ancient as well as modern populations shows that the current inhabitants of Southwest Asia and especially the Levant did not evolve from a single ancestral population. Throughout the Last Glacial Period and up until the start of the Neolithic Period around 12,000 years ago, human foragers who left the African continent experienced severe isolation in various locations in Southwest Asia that led to recognizable genetic signatures specific to these isolated populations (genetic drift).

Our work on the Y chromosome showed that certain Y chromosome signatures are distinctively more abundant in specific populations that occupy the modern-day Levant. These are most likely remnants of population isolation. Subsequent DNA work on the entire genome of modern populations of Southwest Asia as well as on ancient DNA from the same region supports the Y chromosome findings, showing distinctive, recognizable genetic signatures for some of these early isolated populations.

Successive migrations into the Levant and Arabia are manifested by genetic differences, and in some instances, populations occupying not too distant geographic places show a stark contrast in their genetic signatures. Every wave of expansion, coming from different regions, brought different genetic signatures into the population.

What applies to the populations of the Levant applies everywhere else around the globe. People have been constantly on the move since they became bipedal. They just arrived in the Levant much sooner than elsewhere and established settlements much earlier. Recognizable genetic features have been identified for many world populations.

THE MATERNAL AND PATERNAL LINEAGES

Do Y chromosome and mitochondrial DNA patterns in Southwest Asia show any differences? And if these differences exist, do they provide distinct insights into human expansions into the Levant? Do distinct genetic patterns between male and female lineages reveal social behaviors that are specific to the Levant and not elsewhere?

Human mobility, on a scale that could impact the genetic makeup of a population, is usually triggered by a major event, like a war, natural disaster, or abrupt and severe climate change. Some of these events have been well documented, and the Levant has had its fair share of them. While natural events may cause similar mobility between the two sexes, wars have been waged mostly by men. They usually involve massive and prolonged movements of men away from their places of birth, often including the movement of prisoners (also mostly men) and warriors along the way. Alexander the Great marched his army from Greece to the edges of India and occupied a vast empire extending from the Adriatic Sea to the Indus River. The Crusaders, mostly men, came from Europe and occupied the Levant for nearly two hundred years. But variations in paternal lineages (Y chromosomes) have been observed to form restricted geographical clusters compared to maternal lineages (mitochondrial DNA), which tend not to show a localized diversity. This means that women, much more than men, tend to migrate away from their natal household. Whether this male-to-female diversity pattern can be sufficiently explained by the fact that most cultures (70 percent) are patrilocal or by other factors, like more men are killed in wars, remains an open question.

MITOCHONDRIAL SIGNATURES (MATERNAL LINEAGES) OF THE EARLY FARMERS

From a mitochondrial perspective, modern Levantine populations show greater genetic affinity to Europe than do modern Arabian and North African populations. The mitochondrial signatures, or genetic lineages transmitted by mothers to their children, that are common in Europe are also prevalent in the Levant. These shared maternal lineages and their high preponderance in the Levant, compared to their rarity in the rest of Southwest Asia, suggest that these mitochondrial lineages may have spread primarily from a single maternal Levantine source population during the agricultural expansion from the Levant into Europe. After all, Zeus kidnapped Europa from the beaches of Sidon and carried her away to Greece.

Unlike the Levantine populations, the Yemenis, for example, share their mitochondrial signatures with East Africans, while Saudi Arabian maternal lineages are unexpectedly shared with North Africans. One would have expected to see more lineage sharing between Arabia and East Africa, with the latter being a likely source population for the former. In terms of genetic distance (a measure of the proximity of two populations in terms of shared genetic variations), Arabia is much closer to Egypt than it is to Ethiopia. This is not surprising given the long-established trade networks between Arabia and Egypt. The genetic patterns observed are suggestive of a genetic flow characterized by female movement, favored by the patrilocal culture from Arabia into North Africa and in particular Egypt, consistent with the trade routes on the Red Sea that were dominated by Egypt.

None of these shared genetic signatures between Arabia and North Africa are found in Yemen, which has its unique set of mitochondrial lineages, suggesting that population isolation existed in Yemen well before trade routes were established, and some of these populations remained isolated without any genetic exchange with the rest of Arabia. The Yemeni-specific maternal lineages most likely reached Yemen from East Africa and remained highly represented in Yemen. Migration between Yemen and East Africa has been very well documented since the seventh millennium BCE. In addition to trade in incense, spices, and other goods between the old Kingdom of Aksum in Ethiopia, the slave trade constituted a major part of moving people from East Africa to Yemen.

Y CHROMOSOME SIGNATURES (PATERNAL LINEAGES) OF THE EARLY FARMERS

More than 80 percent of the Levantine Y lineage landscape consists of three main paternal lineages (Y chromosome lineages or haplogroups): J1, J2, and E1. This distribution equates with low male lineage diversity and recent geographical presence. These lineages must have arrived in the Levant soon after the Last Glacial Period and not earlier. They reached the Levant after long periods of isolation that led to their low genetic diversity.

The early sedentary cultures formed tight-knit communities that stuck together, sharing resources, which made it easier to withstand food shortages during extended harsh climate periods. J2 and E1 are most abundant across the Levantine coastal areas, while J1 is more abundant in the Levantine hinterland. The decreasing frequencies of J2 and the E1b1b1 subclade (a subbranch of E1) in the inland populations and the corresponding increase in J1 frequency indicate a different source population of these lineages as well as different timing for their arrival in the Levant. Communities in various geographical locations also expanded differently through time and space.

J1, J1-P58 (also known as J1e), and J2 have indeed been observed in the Levant since the Last Glacial Period. J1 and J2 are sister haplogroups that have derived or split from their ancestral parent J lineage—two twigs from the same branch.

THE J ORIGINS

The J1 lineage is the oldest of the three branches—its presence is documented among communities living in the Caucasus and eastern Anatolia during the Upper Paleolithic Period (40,000 years ago). These were the first communities to have an established presence in the Caucasus and Central Asia since *Homo sapiens* left the African continent.

The J2 lineage is believed to have originated a few thousand years later than the J1, somewhere in the northern Levant or southern Anatolia, toward the end of the Last Glacial Period. J1 most likely originated in a location around the Georgian coast of the Black Sea and expanded to the Caucasus by 11,000 years ago, while J2 originated in a southern Anatolian location and reached the Caucasus a little later than J1. Archaeologically, these dates coincide with the change from semi-mobile hunting-gathering communities to the early semi-sedentary settlements in the western Fertile Crescent and northern Levant detailed in previous chapters. They are consistent with the expansion patterns described earlier and correlate with the transition to the early Natufian culture in the Levant. Arboreal areas increased and steppes virtually disappeared, according to pollen data, providing abundant resources that allowed some human populations to expand and occupy semipermanent settlements.

After the initial expansion of the J lineages across Anatolia, the Caucasus, and Southwest Asia, the climate shifted drastically with the arrival of the Younger Dryas (between 12,900 and 11,600 years ago). Rainfall decreased dramatically, and population expansion significantly declined. The Natufian population, which had thrived in the Levant prior to the Younger Dryas, sustained a significant dip and persisted only by abandoning their original dwelling sites and moving into the northern Levant, the Arabian Peninsula, and Yemen, and perhaps Africa.

This period is marked by a restricted movement of the J1, J2, and J1-P58 lineages and correlates with the transition out of the Late Natufian culture, increased sedentism, and rapid population growth in the upper Tigris and Euphrates. Between 10,500 and 10,200 years ago, too, appeared the earliest signs of plant cultivation and, slightly later, of domesticated emmer and einkorn wheat in and near the upper Euphrates Valley.

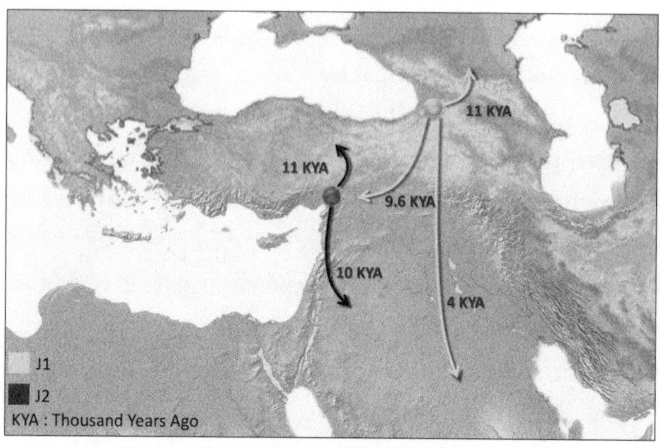

Proposed expansion of the J haplogroups post–Last Glacial Period

COASTAL-INLAND GENETIC CONTRAST

A contrast in genetic diversity between the coast and the inland is apparent in the current Levantine Y chromosomal landscape. This diversity may have resulted from the input of successive migrations from east and west, with each wave consistently strengthening the overall diversity through the introduction of distinct Y lineages. Such heterogeneity or differentiation within a confined geographical region is remarkable. What factors could have contributed to it?

Currently, the J1 lineage is predominant in the eastern Levant, along the Damascus plains, in the Great Rift Valley, and in Arabia. Its distribution coincides with population expansion from the Caucasus, the Zagros Mountains, and eastern Anatolia into neighboring regions along the coast of the Persian Gulf and into Arabia. The highest genetic diversity within J1 is observed in eastern Anatolia, the likely early site of this lineage. The onset of the warmer and wetter conditions around 10,000 years ago, corresponding to the Early Neolithic Period, seems to be associated with a second major period of expansion of the J haplogroups. Around 9,000 years ago J1 expanded farther south and east, and its subbranch, J1-P58, expanded into the Levantine Corridor. J1-P58 appears to have followed a more westerly route around 4,000 years ago. Rather than settling in the Levant, J1-P58 moved farther east through Jordan to Arabia and the Persian Gulf, then up the Tigris-Euphrates, reaching Iran, southeastern Turkey, and Armenia.

J2 reached the southern Levant around 10,600 years ago, which coincides with the appearance of the Harifian (Late Natufian) culture there around 10,500 years ago. The eastern Mediterranean populations carrying J2 lineages expanded earlier than others, settling in regions near modern Lebanon. Their movement patterns differentiate the modern populations currently inhabiting Turkey and Syria from the rest of the eastern Mediterranean populations. Subsequent expansions of this lineage and other genomic signatures differentiate eastern populations (of Armenia, the Caucasus, Iran, and southeastern Turkey) from the other Levantine populations (of Lebanon, Jordan, and Palestine).

All the genetic evidence so far points to a large-scale spread of J2 into Eurasia, eastern Anatolia, and the Caucasus. Its presence in Europe can be explained only through a spread from West Asia, perhaps starting with the Neolithic expansion from the western Fertile Crescent. The high diversity of this lineage observed in the coastal Levant and its spread to Europe from the eastern Mediterranean confirm the presence of this lineage without interruption along the eastern coast of the Mediterranean from at least the Neolithic Period.

The western Fertile Crescent, characterized by a high prevalence of the J2 haplogroup, has been inhabited for a very long time, documented in the region's earliest known cities. The Levantine interior has been and continues to be more arid, with current desert or semi-desert conditions, and has a lower population density. This demographic pattern suggests a reduced level of genetic diversity inland. The current landscape of the region has not changed much since it was initially inhabited.

Attempting to reconstruct, solely through genetic data, details of when and from where ancient populations moved and which migration paths they followed is an insurmountable challenge. To gain a more comprehensive understanding of the genetic data, we must supplement it with historical, social, archaeological, and climatic information. Equally vital, we must visit the populations being genetically investigated to learn not only about their cultures but about their personal narratives and histories—crucial and often necessary for drawing meaningful conclusions about their ancestors.

Populations cannot be distilled into a few genetic markers like lineages and haplogroups. These terms, while scientifically significant, can reveal only part of the large and complex puzzle of human migration. They serve as essential pieces but remain devoid of meaning when considered in isolation.

CLIMATE AND POPULATION DYNAMICS

Genetic diversity, the degree of genetic exchange in a given population, is noticeably higher on the Levantine coast compared to the interior. The higher coastal diversity may be the result of successive migrations from the Anatolian plains and the Caucasus after the Last Glacial Period. The coastal inhabitants, unlike their inland neighbors, did not experience genetic drift. J1 remained isolated in inland populations for a long period of time. There was no evidence to support the genetic drift hypothesis, and we have no reason to believe that the populations that migrated toward the inland were much smaller or more genetically homogeneous than the ones that migrated toward the coast.

Besides, the high coastal J2 frequency and the amount of genetic variability observed within this haplogroup indicate a continuous, uninterrupted occurrence of this haplogroup along the eastern coast of the Mediterranean. Therefore the difference between the coastal and inland populations is best explained by successive waves of migration into the Levant after the Last Glacial Period, namely those Anatolian and Iranian farmers. The spread and distribution of the J1 and J2 lineages resulted in a male phylogenetic landscape that correlates with the movement of sheep and goat herders expanding south from the Zagros Mountains and eastern Anatolia during the Neolithic Period.

Furthermore, the J1 and J2 haplogroup distributions across this region align nicely with the historical patterns of rainfall during the Agricultural Revolution in the Fertile

Crescent. Agricultural communities that settled along the coasts were rich in J2 lineage, while J1 seems to have become prevalent among the herding populations that occupied the inland. Although the geographical gradient of J1 supports expansions from earlier refugia, certain J1 subgroups (J1-P58) seem to have originated in the Arabian Peninsula and were likely introduced into the Levant more recently, possibly during the Islamic expansion.

THE E LINEAGE

The geographical distribution and variability of haplogroup E1b1b1 resulted from back-and-forth migration between the Levant and North Africa. This haplogroup exhibits its highest frequency in North Africa, particularly in Egypt, Morocco, and Tunisia. It gradually diminishes as it extends beyond North Africa through the Horn of Africa and the Levantine Corridor into the Arabian Peninsula and Central Asia. The prevalence of E1b1b1 also decreases in Spain and Southern Europe. The Islamic conquest of Iberia was most likely behind the spread of E1b1b1 to Iberia through the Strait of Gibraltar. This conquest involved armies with a significant number of Berber recruits, whose presence began with the Umayyad Caliphate (661–750 CE) and persisted for several centuries. A distinction between Arabic and Berber pools of E1b1b1 in North Africa was emphasized in a study involving populations from Morocco, Tunisia, and Algeria.

The E1b1b1 diversity patterns that are noted in the Levant and North Africa are consistent with early migrations (Neolithic) from the Levant to North Africa. These early migrations were followed by several small and time-interrupted gene flows also from the Levant to North Africa, followed by a rapid expansion. Back-migration from North Africa into the Levant is also evident from this haplogroup distribution, likely influenced by forced migrations and climate changes. Overall, the E1b1b1 frequency gradient between the coastal and inland Levant suggests an origin and expansion of E1b1b1 near the coast with limited migration inland.

The genetic makeup of the E1b1b1-M81 branch (subclade) serves to dissect these migration events, revealing a geographical separation of this branch between North African (Moroccan and Tunisian) and Levantine populations. The number of mutations along this branch offers a time estimate for the most recent common ancestor in North Africa, aligning with a potential Neolithic origin.

FROM CLIMATIC DISPERSAL TO EARLY SETTLEMENTS

The patterns of the J lineage distribution are consistent with the fact that Southwest Asia, for significant periods, was generally uninhabitable. During the Last Glacial Period, populations with J lineages would have experienced a range of habitats with significant population size fluctuation over time, with the Caucasus showing some microclimatic variations that most certainly would have impacted living conditions. The apparent geographical structure of dispersal associated with the J lineages may indicate the early appearance of patrilineal descent (descent traced only through the paternal line) and patrilocal patterns of dispersal (movement that is paternally dominated), which tend to be associated with the later Neolithic expansions.

During much of the Last Glacial Period, the general aridity of the Levant, except during moist periods, likely barred expansion farther south into the region. The Early Holocene Climatic Optimum (about 10,000 years ago) roughly corresponds to the earliest expansion dates for Levantine population differentiation.

These expansion dates correspond to the prepottery Neolithic Period, when archaeological evidence from megalithic ritual sites (like Göbekli Tepe, one of the earliest sites of large stone construction) indicates increasing social complexity. They also coincide with the development of wider-ranging obsidian trade networks, spreading this volcanic glass, an important material for making cutting tools, to Anatolian sites and Jericho. A substantial trade network, using diverse stone-tool manufacturing technologies and styles, ranged from the Levant to southern Iraq. Given these ancient artifacts, the genetic record suggests that the populations involved with Göbekli Tepe likely participated in these early expansions into the Levant and the rest of Southwest Asia following the paths marked by the J lineages, confirming the genetic findings discussed above.

A major deterioration in the climate that occurred at approximately 8,200 years ago would have caused another retraction of populations into local refugia and resulted in greater population density in these regions. The final major period of expansion and differentiation observed in the J lineages in Southwest Asia occurred around 7,000 years ago, not long after another warming trend resulted in sea levels rising by as much as thirty-five meters around the eastern Mediterranean. Interestingly, the expansion of the J2 sublineage J2a2 immediately followed this event. The relatively late expansions of J2a2 correspond with possible expansion routes unconstrained by arid regions, suggesting multiple established destination settlements and routes already in place with relatively strong population expansions. The earliest Neolithic sites beyond the central plains of Anatolia, such as Hacilar, appear to contain rare evidence of pottery in the earliest layers. This supports the suggestion that the earliest westward expansions from southern Anatolia may have occurred at the end of the prepottery period or just after, in the early Pottery Neolithic.

Around 7,500 years ago both J1 and J2 expanded into the Levantine Corridor, coinciding with improved climatic conditions. This time period marks the new Pottery Neolithic Yarmukian culture. It also corresponds to marked differentiation between populations that occupied the southern Levant and those in the northern part where the Halaf culture flourished. This separation was marked by climate differences that led to distinct agricultural methods. In the north, rain was abundant, and farmers could rely on its consistent occurrence, while the dry south needed artificial irrigation and was more developed.

The time correlations of these genetic reconstructions with climate, archaeology, and history appear to be reflected in the archaeological record at Göbekli Tepe, in the obsidian trade, and in the settlement of the Fertile Crescent following the dispersal routes of specific genetic signatures. However, populations at Göbekli Tepe clearly traded widely and perhaps had some opportunity for admixtures. Ancient attestations, much later than at Göbekli Tepe, of languages now extinct that are deemed linguistic isolates reveal that the populations had an even more complex structure than accounted for by genetics. But genetic signatures are not populations, and these population expansions represent rough average reconstructions, given the level of resolution applied for populations, regions, and lineages.

The Neolithic Revolution did not replace earlier cultures that had wandered across Southwest Asia. It brought them together, from their isolated refugia, and expanded their presence throughout this region and beyond. These expansions led to major population growth, the development of large-scale farming industries, irrigation technologies, the spread of Indo-European languages, and the creation of the first cities, all of which gave rise to the early civilizations and their cultures. It is a remarkable story of struggle and continuity that brought people together, a history that should be well contemplated and cherished.

Our ancestors faced near extinction on numerous occasions, and getting to where we are today was a tenuous and hard slog. Climate fluctuation was the main challenge for the first inhabitants of the planet, even as it continues to be the major challenge for humanity today.

CHAPTER 10

What Is an Indigenous Population?

To describe a population,
gather firsthand information,
otherwise steer clear

BEFORE WE DELVE INTO THE populations and histories of the Levant, it is critical to clarify the notion of indigenous populations.

LOST POPULATIONS

In 2005 my colleagues and I traveled on an expedition to Chad to investigate two "indigenous" groups of people and to obtain DNA from them for phylogenetic analyses. The first population consisted of a few hundred individuals living mainly in two villages (Gori and Damtar) along the Chari River in the sub-Saharan, southern part of Chad known as the Moyen-Chari region. These people are called the Laal, in reference to the unclassified language they speak, called Laal, that uses clicks for paralinguistic purposes (nonverbal gestures for communication).

The second population we studied was a group of nomadic tribes called the Toubou, living in the Tibesti Mountains in the extreme north of Chad. These people speak a

Nilo-Saharan language and are known to be the "African Gatekeepers," the tribes that guard the northern gates of Africa. Our plan was to rendezvous with local experts including guides, drivers, and a cook, with whom we planned the logistics of our trip. We were first to drive south to try to locate the Laal speakers, then drive north to the Tibesti for the Toubou tribes.

On a sunny October afternoon, our Air France Airbus A320 touched down on the single short runway of N'Djamena Airport, in the capital of the Republic of Chad. Exiting the plane, we followed the crowd of passengers into the narrow corridor leading to passport control. It was hot, humid, and very crowded. Our driver delivered us to a hotel on the edge of the Chari River, the main source of fresh water and fish for the local population.

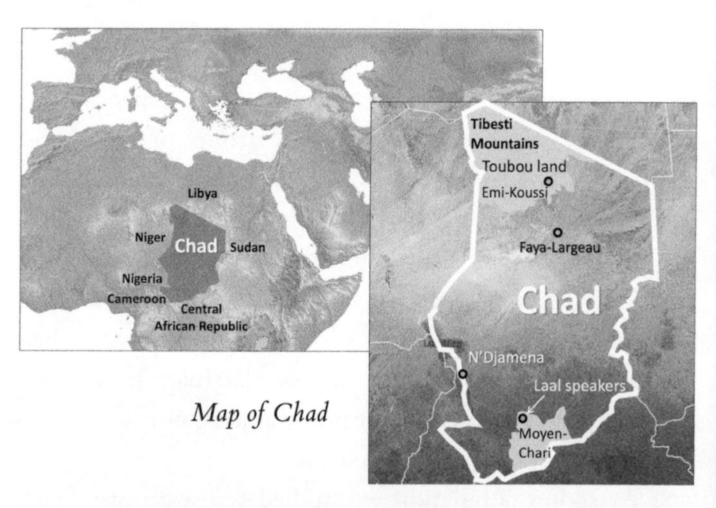

Map of Chad

After a weeklong delay to obtain all the necessary permits, we were finally ready for the trip. We started our two-day journey south in three four-wheel drive vehicles loaded with food, water, and other supplies. In addition to what the cook brought, we loaded up on dried food—mostly beef jerky, peanuts, and candy—and lots of bottled water. We all had our antimalarial drugs and some basic over-the-counter medications, in addition to antibiotics and iodine capsules for water disinfection.

After several short stops, we arrived at our destination. The last stretch was particularly rough as we had to drive through thick and very muddy sub-Saharan forest, totally unmarked and uncharted. At some point, one of the drivers wanted to run over a guinea fowl that crossed our path to roast for dinner; luckily she outsmarted him. As we drew near, we could see dense forest mist and some white smoke in the distance. As we got closer to the smoke, a small clearing appeared, dotted with round green huts with brownish cone-shaped roofs. The hut bases were made of forest wood and covered with tree leaves and green-brownish shrubs. Several small children came running toward us while their watchful mothers, most with babies in their arms, kept their distance. As we were setting our camp for the night not far away, we had a visit from the male elders of the community. I counted a total of fourteen men and four youths.

None of the men were above the age of fifty from what I could tell. Their leader, who did not appear to be the oldest among them, greeted us. Our guide, from the closest town, spoke a different Nilo-Saharan dialect but could somewhat communicate with them. We all sat together around a bonfire that they had prepared and lit for us. They brought fresh water and smoked fish that must have come from the Chari

The males of the Laal-speaking tribe we visited in Chad: fourteen adult males and four young men

River, burbling softly close to where we were. We shared our food with them and spent a few hours under a perfectly and naturally illuminated sky, away from any form of civilization, discussing the secrets of life and humanity's existence without uttering a comprehensible word to one another. We were mesmerized by their kindness, their serenity, and most of all their contentment. They live off the Chari River, they farm the land, and they trade with similar sized communities, also living in seclusion, farther up the river. Our guide explained the reason for our visit. They appeared to be intrigued and happily provided us with the DNA—saliva swab samples—that we were after. At no point did the women or little girls come close to us.

The next day we woke up very early to forest sounds. The morning was green, crisp, and fresh. For a long moment, I felt envious of their way of life in the middle of this thick and remote forest.

Our morning discussion was disrupted by the sound of a nasty cough, coming from a child in the distance. He seemed to be about six years old, observing us from behind a tree. He was a cute little boy, wearing a small piece of cloth around his waist. I waved at him and gestured to him to come closer. A colleague went to try to find some biscuits or chocolate to give him but did not find any.

As the boy approached, we could sense some hesitation in his walk, so our youngest colleague moved toward him and extended his hand. The boy reached out and tapped him on the palm in a playful way and giggled. During our brief interaction with him, he kept on coughing this disturbing cough that I remember to this day. We debated whether to give him some of the antibiotics that we had brought with us, but we were afraid to do so, and after mulling it over, we decided against it. None of us was a physician, and we were not sure if his cough was caused by a bacterial infection or not. We did not want to risk giving him an antibiotic that might end up harming him. I vividly remember this cute little boy, his shy smile, and the nasty cough he had, and I keep asking myself to this day whether we had made the right decision by not giving him the antibiotics. It pains me that I will never know what happened to him.

Slowly, as the sun rose and the forest dew disappeared, the men from the night before appeared among us. Our cook had prepared breakfast and coffee. They tasted some of the coffee and seemed to like it. As we readied to depart with heavy hearts, I thought about how a cough that could be easily taken care of at home could be deadly in this community. Was this the reason that it had no elders? I was perplexed.

Around midday we left these fourteen men and their families, who appeared to be living undisturbed and serene in a tropical forest far from modern cultures. They must have lived in almost complete isolation for thousands of years. They probably had a genetic signature that distinguished them from their neighboring communities, we thought, and we were excited to find out.

WERE THEY REALLY INDIGENOUS?

Several weeks later we analyzed their DNA in the lab and observed to our great surprise that they may have had contact with populations who lived as far away as Asia Minor and/or Southwest Asia. This contact, which must have occurred between 5,000 and 7,000 years ago based on our analyses, may have been due to one or more back-migration events to Africa.

Modern humans adopted forms of primitive farming in West Asia and Asia Minor soon after the Last Glacial Period. Before the major population expansions from Southwest Asia and Asia Minor, people still lived in small groups as hunter-gatherers. The expansions occurred during a period when the climate in North Africa was more favorable and more humid than the climate in Southwest Asia. During the African Humid Period (between 20,000 and 6,000 years ago) North Africa was dotted with lakes and lush with vegetation. The African Humid Period resulted from the tilt of the earth's rotational axis—which causes gravitational changes between the earth and the moon and occurs once every 20,000 years or so—transforming the arid North African landscape into a savannah.

This lush environment attracted people in Southwest Asia to migrate back to Africa and beyond in search of fertile land. We believe that during the later years of the African Humid Period, the Laal speakers and people from Southwest or Central Asia might have had contact. It was the last time the Laal speakers interacted with anyone outside their closed community and before they went back to almost complete isolation.

So are the Laal truly indigenous? If we were to consider genetics only, they appear to be admixed with populations from Asia. But the last admixture happened at least 6,000 years ago, and since then they have been living in complete seclusion. How far back in time do we have to go in order to call a population indigenous from a genetics or anthropological perspective? But does it really matter? The Laal are indigenous irrespective of how and when they last admixed.

THE ROAD TO THE TIBESTI MOUNTAINS

We paused for a few days in N'Djamena, where we extracted the DNA that we had collected from the Laal population. We used a makeshift DNA extraction station that we installed at the University of N'Djamena. Then we loaded the cars and set out at five in the morning, heading north. Our destination was the Tibesti Mountains on the northern tip of Chad, where the Toubou tribes live.

First we would drive to Faya-Largeau, the last big town in northern Chad that can be reached by car. We had to cover a distance of approximately eight hundred kilometers, and considering the rough road conditions, the absence of designated roads in some places, and the terrain, we estimated we could drive no more than three hundred kilometers per day. The guide and drivers were adamant that we should stop driving at dusk, as it was way too dangerous to drive at night. Once we were in Faya-Largeau, we had planned to take a small Cessna to the Tibesti Mountains. At the time of our expedition and still to this date, the Tibesti are controlled by Chadian rebels, mainly composed of Toubou, fighting the government. We had hired a pilot to meet us at Faya-Largeau and take us to the Tibesti. Afterward he would fly us back to N'Djamena, while the drivers and the guides would return in the cars—that was the plan, anyway.

That first day we drove about sixty kilometers on dirt roads, but at midday the view became foggy. Yacoub, our lead driver, pulled over, got out of the car, and conferred with the other drivers and our main guide. They informed us that this was not fog—we were in the middle of a dust storm, and we could not drive through it. The guide could not see far enough ahead to navigate through the desert. We had seen no designated roads or any signs of a road since we left a cheese farm where our cook had had us stop.

At around noon, it was still dark. We decided to set up camp and wait in the hope that the dust would clear the next day. We pitched our camp and had lunch. The main dish was the artisanal cheese that the cook bought from the farm earlier in the day. I decided to skip the cheese, as it did not look too appetizing to me, but my colleagues had their fill and appeared to enjoy it very much. I opted for some tasty beef jerky and a few green olives instead.

At about seven P.M. we decided to sleep as we had to rise very early to try to make up some of the lost time. As the guide and the drivers were setting up the tents for us, we noticed they were talking animatedly and appeared to be worried. Yacoub asked us to follow him, and with his flashlight he pointed to the ground, where we could see something moving in the sand. At a closer look, the entire ground where we were to pitch the tents was infested with yellowish-brown, agitated scorpions, each the size of a large thumb. Apparently, this was common in this area at this time of year. We contemplated sleeping in the cars, but there were too many of us to fit comfortably. After a long debate, we finally agreed to sleep on the small sponge mattresses that each of us had in our individual tents. These mattresses were thick enough that a scorpion tail could not strike through it, or so the guide assured us.

I did not close an eye that night for fear of getting stung by one of the dozens of scorpions lurking under my mattress. My colleagues did not sleep either, but for a totally different reason. They were as sick as dogs from the cheese they had devoured a few hours earlier. They vomited for hours on end. After the vomiting came diarrhea—and we were in the middle of nowhere, stuck in a sandstorm. Despite the fact that we had brought enough water to last us five days for a three-day journey, dehydration was our main concern. Now we were going to need much more water than expected, especially since we had lost one travel day already.

After a long painful night, we resumed our slow progress north. The driving conditions were as horrible as the previous day if not worse, and we had to pause many times for my sick colleagues. At about two P.M. we stopped for the day and pitched our second camp of the journey. Despite the inspiring confidence of our guide and drivers, I was certain that we were lost in the Chadian desert. Whatever antibiotics we had were used up, and we did not have enough to go around for a full course for anyone. Luckily, I had also brought some metronidazole pills—a potent antibiotic used to treat anaerobic bacteria as well as some parasites. We divided these pills equally among our sick colleagues. It was already day two, and we were not nearly a third of the way into our journey.

As we settled into camp, the guide approached us looking seriously worried again. We would run out of water tomorrow, he warned, and we did not have enough to last for the duration of the trip. This was impossible, we objected. We had brought enough for five days as a precaution, and we were only in day two, so we must have more than enough. When he insisted and showed us our water supply, we realized the gravity of the situation and panicked. How could this have happened?

We immediately decided to go on water rations, but my colleagues who were fighting off food poisoning needed to be constantly hydrated. After tallying what we had left, we calculated that if each of us drank one liter of water a day, we would have enough for three days, which according to our guide would be enough to get us to Faya. He told us that he might get us there in less than three days if the storm subsided. The next day things did not get any better with regard to water consumption. We started putting our names on the water bottles. That did not help, and by the end of the day we were totally out of bottled water.

Under severe stress, an egotistic survival mode kicks in and people who under normal conditions behave rationally and civilly assume bizarre and selfish behavior. Those who had been friends the night before and who have known each other for decades become total strangers. We had a hard time observing what was happening to us and were shocked to realize how our behavior as humans could be so frail and fragile under extreme conditions.

Nor were the road conditions getting any better. The sandstorm continued unabated, and we needed drinking water. We had brought with us a tank of water to use for washing dishes—it was unsafe to drink. We had no choice now but to tap into it and try to satisfy our needs. We boiled small pots of it, then dissolved a few tablets of iodine that we'd brought with us. In retrospect, I think we overdid it; we added too much iodine to the water. We had no water filters, so we had to drink it unfiltered, still yellowish from the iodine. It tasted awful, but it served its purpose in limiting water consumption to a minimum. We finally reached Faya at seven P.M., six days after we'd left N'Djamena.

A TOWN IN THE TIBESTI

Faya is an oasis amid the rugged and harsh Tibesti Mountains. It is the land of the Toubou, nomadic tribal warriors who have lived across the North African corridor that connects Africa to Southwest Asia and controlled the sub-Saharan African gates for thousands of years. They claim ancestry in Africa as well as Yemen. They have their own Nilo-Saharan dialect and are divided into two major groups, the Teda to the north and the Daza to the south. We had learned about them and their culture before we arrived in the Tibesti.

We met our contacts in Faya and arranged to meet some Toubou men in the area. With trepidation we drove for a few hours out of Faya for our first meeting with these mysterious and fearless warriors who have managed to live in one of the most hostile and inhospitable places on the face of the earth, and for so long. When we arrived, we were greeted affably by a group of men who had very distinctive facial features. We got to spend several days with these welcoming people and learned more about their history and their way of life. Importantly, we were able to obtain samples for our DNA work.

THE TOUBOU TRIBE

While we were collecting DNA from the Toubou tribe in the Tibesti, they told us stories about fair-skinned men from Europe who used to come and live among the Toubou and who left many descendants. These descendants had been totally assimilated into the host population and could no longer be identified. In fact, it takes only a few generations for traits that are distinct from the host population to vanish from sight.

Indeed, a Europe-specific genetic mutation that is associated with lactase persistence was present in about 2 percent of the Toubou. This mutation was totally absent from all the other Chadian populations. Lactase is the main enzyme that digests lactose (the sugar that's found in milk). In humans and other mammals, the production of lactase slows significantly after the weaning period. But certain mutations in the gene that codes for lactase lead to the continuous production of the enzyme, resulting in what is referred to as lactase persistence.

Mutations that cause lactase persistence are thought to have been fixed in populations after they domesticated animals and adopted dairying practices around 9,000 years ago. The earliest dairy farmers lived in the Anatolian plains and Mesopotamia and brought their practices to Europe 7,000 to 6,000 years ago. A specific mutation in the lactase gene was highly selected for in early European farming populations, for whom bovine milk consumption constituted a major portion of their diet. The story of lactase persistence and how humans adapted to milk consumption is more complicated than a mutation in a gene. It has to do with other genetic adaptations, with causes still unknown, and most important with how the thousands of bacterial strains (the microbiome) that colonize the human gut reacted to nonhuman milk and dairy consumption.

A genetic variant that contributes significantly to blue eye color was also observed among the Toubou. Whether these markers came from the Europeans who settled among them in the last couple of centuries or from a much earlier interbreeding event following the Eurasian back-migration into East Africa has yet to be fully resolved. What is certain, however, is that the Toubou have significant European ancestry.

In fact, the DNA analyses showed that, genetically, the Toubou have both African and Eurasian ancestries. So how and when did they get to the Tibesti in the first place?

When modern humans left Africa some 200,000 years ago in search of more hospitable places, they settled in and around the Levant. When climate conditions favored Africa again around 20,000 to 18,000 years ago, some tribes including the Toubou may have migrated and settled in and around the North African migratory corridor and established their nomadic communities. Their DNA revealed that they had come into contact with Eurasian farmers about 8,000 years ago when these farmers migrated back into Africa, following the green corridor created during the peak of the African Humid Period. This corridor witnessed an intermittent, very active back-and-forth flow of people between Eurasia and Africa.

But how did the Toubou survive for so long in this hostile environment? Between 20,000 and 8,000 years ago, the climate in North Africa was somewhat favorable because of the periodic orbital eccentricity that brings the earth closer to the sun. This period of warmth brought warm, moist air from the oceans and seas that, when cooled inland, turned into precipitation. North Africa got a fair share of that moisture. When this humid period ended, large parts of North Africa reverted to an arid, harsh landscape like the Tibesti, but the Toubou warriors managed to stay put through it all. They are indeed Africa's gatekeepers, the guards of the Tibesti. While the genetic background of these warriors is heavily admixed with DNA that resembles modern Europeans, they, like the Laal population, have been mostly isolated, occupying these lands for 20,000 years.

THE JOURNEY BACK

Back at our base camp in Faya, feeling proud that we had accomplished our mission with success, all I could think about was the trip back to N'Djamena. Although anxious, I was somewhat relieved that we did not have to make the trip by car.

Our Cessna had arrived the night before. We said our goodbyes to our guide, the drivers, and the local colleagues who came with us on the trip from N'Djamena and wished them luck in their journey back. We took the plane very early the next morning, heading for the Chadian capital. It was still dark; the sky was clear, but the air was very crisp. During the flight, I was sitting right behind our South African pilot, and I could not help but notice a huge scar extending from his upper right arm to his neck. I was curious but not sure I wanted to ask, especially since the small plane was shaking a bit too much for any of us to be comfortable. I finally gathered enough courage and asked him: "Were you a military pilot before doing this business?" He smiled and pointed to his scar and said: "Do you really want to know?"

"No," I said. All I wanted then was to make it back home in one piece.

INDIGENOUS OR ISOLATED?

Indigenous populations, like the Laal speakers of sub-Saharan Chad, or the Toubou of the Tibesti Mountains, who have been isolated for such a long time without interbreeding with outside populations are expected to have less genetic variation. Nearly 50 percent of the Y chromosome signatures of the Laal population, which has been in almost complete isolation for many centuries, are not African but have been introduced through back-migration from Asia Minor or Central Asia. The Laal population is an example of a perfectly "indigenous" population with its own "indigenous" language; the only access to it is by driving through a thick forest for two days on unmarked or undesignated roads using four-by-four vehicles with multiple guides, only to find out that nearly half of their genetic ancestry is derived from back-migration.

The Toubou tribes have up to 30 percent European ancestry. Other indigenous populations may be less interbred because of their longer isolation periods without interbreeding with external populations. In the Americas, for example, the Native American populations that left Beringia more than 10,000 years ago spread throughout the Americas without outgroup interbreeding and developed into subgroups that can still be genetically identified.

A DNA SAMPLE DOES NOT COME FROM A FREEZER

In telling this adventure, I wanted to point out that a DNA sample does not just come out of a freezer somewhere. It comes from people, who have fascinating stories and histories. The more we learn about the people who provide us with their DNA, the better and more accurate stories we can write about them and their histories. What we learned from the people we interacted with was not only enriching and gratifying on a personal level but was crucial to us in interpreting our DNA findings. Our conclusions would most likely have been drastically different and perhaps less valid had we not visited and interacted with them, experienced their harsh environment, and learned from and about the populations. Making interpretations and drawing conclusions about indigenous populations based on analyses conducted on "out of the freezer" DNA samples without interacting with them can be very misleading and, most often, inaccurate.

PART V

HUMAN MOBILITY and CULTURES

CHAPTER II

The Early Dynasties and Empires

Empires rise, but what remains of them after they fall should be cherished

WHO CARES ABOUT ARTIFACTS?

Destroying a culture is destroying a common heritage for all humanity. The Levant, like many parts of the world, has witnessed and is still witnessing large-scale destruction of important artifacts that tell stories about the many cultures that have existed since the first human community was established there.

Throughout history, cultures have systematically destroyed ancient archaeological remains and effaced traces of older cultures. Inert ancient carved stones and other artifacts must have enough influence and dominance to become targets of destruction. Their influence is in what they represent, in the stories they tell, and in the cultures they preserve. Obliterating a slab of stone or a statue aims not merely to destroy precious monuments; much more, it aims to erase an entire culture and with it an identity.

So lie in peace, Sargon; no one will know you ever existed. After all, Nineveh sounds like a city from *Star Trek.*

Why should anyone care about a small city that was founded more than 9,000 years ago on the banks of the Tigris River? A place where people first learned how to live together in communities by farming the land, domesticating animals, and caring for one another? These people spoke languages that have long disappeared, etched artwork on rocks, and believed in Ishtar, a strange goddess of love, sex, and beauty. They called themselves Akkadians, Sumerians, Babylonians, and Assyrians. We can't keep track of who came first or after, and so who cares? Why should we care about a distant past? Who cares about a few limestone sculptures that were recently barbarously destroyed? They were not the first to be destroyed and certainly will not be the last.

BUT WE SHOULD CARE ABOUT ARTIFACTS

Those who painstakingly etched these stones day in and day out for years sought immortality, a testimony to their gods, to future generations, to their existence and our own. These artifacts do not belong to a group or a country. They belong to us all, to humanity. They constitute our common heritage. They are part of our collective memory that we must not only cherish but protect with vehemence. Our heritage makes the very fabric of our identity. After all, it is this connection to our past, tying us all together, that we need to protect.

AGRICULTURE CHANGED EVERYTHING

The adoption of agriculture was a giant "evolutionary" leap. Jared Diamond describes it as "a watershed moment for the human race . . . our greatest blunder." In his mind and that of many others, this shift in the lifestyle of hunter-gatherers, while tremendously beneficial for early communities, over the long run ushered in a certain laziness that resulted from the relative abundance of food and from the storage habits that made access to food easy.

Farming and domestication led to behavioral changes, and as these farming communities grew in space and population size, they became the target of less affluent communities, and this resulted in different kinds of struggles.

In her book *The Egalitarians: Human and Chimpanzee*, published in 1991, Margaret Power describes how one of our closest primate relatives, the chimps, became aggressive when researchers provided them with designated feeding places. They started to compete for the easily accessible bananas, when only a few days earlier they were collectively searching the forests for wild fruit trees to eat from.

With agriculture and farming came envy, aggression, protective walls, and conquests. These manifestations brought social stratification and more complex structured communities that witnessed the budding of rulers and power struggles that expanded beyond the local burgeoning farming community. As communities flourished, the conquests grew in scale and became associated with wars involving large-scale human mobility and, with it, more genetic admixtures.

The original communities that enjoyed some degree of isolation ceased to exist and were replaced by larger and very often admixed communities, a nightmare for population geneticists.

HOW DID THE EARLY COMMUNITIES AND THEIR CULTURES EVOLVE?

The small, scattered populations that occupied the Levant after the establishment of the first city-state in Sumer were numerous. Some are better known than others, some remain unknown to us today, and some may never be identified. DNA analyses of modern, living individuals can reveal the origins, and unravel some secrets, of past ancestors. Interpretations about populations of the past can, however, be very tricky, considering not only the level of admixture that has been occurring since the establishment of the first cities but also the fluidity of the populations.

Ancient DNA analyses are helping with modern findings as they provide anchors for comparisons. But without a comprehensive approach that is based both on solid archaeological evidence and on an appreciation of history, even though historical narratives can be unreliable, DNA interpretations can very well be off target or often misleading.

The following narrative is meant to provide a historical perspective about the various populations that occupied the Levant and Mesopotamia since the appearance of the first cultures 6,000 years ago. It is meant to highlight the complexity of the Levant from a population perspective and to underscore the difficulties that are encountered in interpreting DNA findings when trying to reconstruct a history that has witnessed many successive populations. The following succinct and chronological narrative about the Levant shows how rich and complex its history is and what makes it such an interesting and captivating place. It is not meant to be exhaustive, but rather focuses on major historical events that shaped its cultural, geographical, and genetic landscape through the last several millennia.

While Babylon was not technically in the Levant, it played a major part in the region's history and shaped it to the way it is today. When a group of people gather and their conversation is not making any sense, where everyone is shouting at one another at the same time without understanding what anyone is saying, this cacophony of sounds is referred to as a "Tower of Babel," an old expression still used in many parts of the Levant and beyond. The saying is basically referencing the innumerable languages that were spoken in Babylon during the same time period. It was in and around Babylonia that the first cultures appeared and from Babylonia that many cultures radiated.

THE SUMERIANS AND THE START OF TERRITORIAL EXPANSIONS (THE FIRST WARS)

The first Sumerian cities, or city-states, were established toward the end of the fifth millennium BCE (around 6,000 years ago) by the same people who had been dwelling in Mesopotamia since the Last Glacial Period. These were the Ubaidians, and their culture (Ubaid) defined the Uruk Period, when the first cities—Ur, Uruk, and Eridu, among others—were established. The Sumerian culture was the product of the Samarran, Halaf, and Ubaid cultures. They were the first cultures to develop community and urban societal lifestyles and city-state governance. They spoke a language that has long been extinct, and their cuneiform "Sumerian pictographic characters" were recognized as early as 5,100 years ago in the city-state of Uruk.

The land they occupied, between the Tigris River on the east and the Euphrates on the west, was the most fertile at the time, a delta that created an alluvial plain made of silt and other river deposits, perfect for agriculture—an area that is called southern or lower Mesopotamia (*meso,* "in the middle"; *potamia,* "rivers"). There the concept of land management first arose along with the invention of elaborate irrigation systems by channeling river water through canals.

The Sumerians were not left at peace for a long time. Another group of people, called the Akkadians, were roaming Mesopotamia around the same time, but farther to the north. Soon they encountered the Sumerians and eventually brought them to submission. Whether it was by force or intermarriage and a smooth transition that the Akkadians overwhelmed and replaced Sumerian culture is still a matter of debate. What *is* certain is that the Akkadian culture replaced, dissolved, or dominated the Sumerian culture.

THE AKKADIANS

The origin of the Akkadians has been disputed: Some claim a northeastern origin (Anatolian plains) while others propose the Caucasus. Still others advance a theory that late in the third millennium BCE they arrived in Mesopotamia from the northern Levant. From a linguistic perspective, Akkadian belongs to the East Semitic languages, and the northern Levantine populations spoke West Semitic at the time.

To be more precise, they spoke Northwest Semitic. Some linguists claim that Eblaite, another East Semitic language, the closest to Akkadian, was also spoken in the northern Levant along with old West Semitic languages.

These linguistic arguments favor a continuous flow of people (and their languages) from the west (Levant) to an already somewhat populated Mesopotamia and not *en masse* migration from northwestern Levant into Mesopotamia consistent with the genetic evidence discussed earlier. Genes and languages sometimes migrate together but never with the same intensity. Languages can easily spread without genes, and genes can spread independently of languages. It may have been a west-east language journey that brought the Semitic language to the Akkadians. At the height of Akkadian rule, the emerging lingua franca was Akkadian, while during the same time period in the Levant, several West-Central Semitic languages were spoken—Phoenician (Canaanite), Aramaic, and ancient Hebrew.

The Akkadian culture arose in northern Mesopotamia around 4,350 years ago and established Agade (Akkad), a city that has been neither rediscovered nor identified. While Akkadians embraced most of the Sumerian habits, rituals, and ways of life, they maintained their Semitic language. Akkadians went beyond Sumer's dominion and conquered the Elamites to the east of Sumer (today's Khuzestan, in Iran) and reached the Levant and Dilmun (today's Bahrain). In less than half a century, Sargon the Great, the first Akkadian ruler, established the first world empire, the Akkadian. Sargon's empire was, however, short-lived.

ASSYRIANS

After Sargon's death, the Akkadian dominion faltered under the persistent and continuous assaults of the Gutians, who invaded Mesopotamia from the Zagros Mountains. Around 4,190 years ago the Akkadian Empire succumbed to the Gutians, who maintained some level of control over some of the city-states in Mesopotamia. Their power was, however, continuously challenged by several city-state kings, and in particular the king of Uruk. The king, who had maintained some level of autonomy throughout the Akkadian rule, declared total independence. This period of city-state rule (c. 4,120 to 3,950 years ago) restored some of the Sumerian reign, but it soon ended with the arrival of the Amorites from Amurru (in present-day Syria). The Amorites were Semitic people who dwelled and dominated part of the northern Levant before they moved into Mesopotamia to establish the first Babylonian dynasty around 3,900 years ago. Soon afterward Mesopotamia became divided into two main cultures: one in the north along the upper Tigris River, called Assyria, with Assur as its capital, and one in the south, called Babylonia and claiming Babylon as its capital.

HAMMURABI'S BABYLONIANS

The Amorites were already in southern Mesopotamia when the Neo-Sumerian dynasties started to fracture. Soon after the Sumerian dynasties fell, the Amorites consolidated their power and established many dynasties along the Eu-

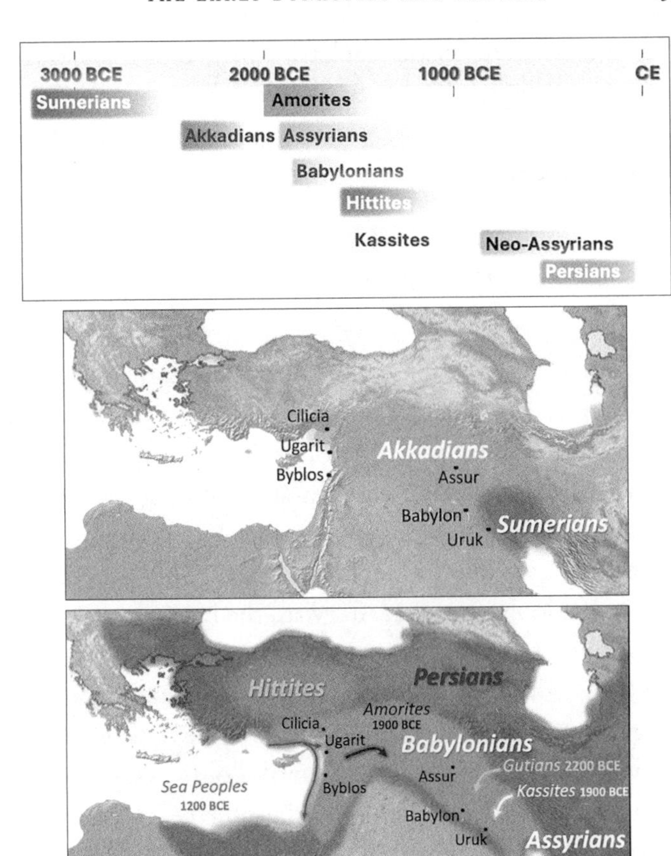

Timeline of the early dynasties

phrates, the most famous and lasting of which was Babylon. The Babylonian dynasty, during the reign of its sixth king, Hammurabi (c. 3,810 to 3,770 years ago), dominated southern Mesopotamia, parts of northern Mesopotamia, and large parts of the Levant. Hammurabi established the first code of law and had his scribes carve it in stone in the Akkadian cuneiform script. Under his rule, Babylonia was a flourishing empire. Today the basalt slab on which Hammurabi's law was inscribed stands in perfect condition at the Louvre Museum in Paris.

THE HITTITES AND THE KASSITES

Around 3,620 years ago Babylonia, under its first dynasty, was invaded by the Indo-European-speaking Hittites of Asia Minor. The Hittites first appeared in central Anatolia (Cilicia) and established Hattusa as their capital around 3,670 years ago. They expanded southward, invading the Levant and engaging with the Egyptians who had been enjoying supremacy over the Levant unmolested. This occupation did not last long, and soon afterward the Hittites went back to Anatolia. Then the Kassites from the east (Zagros Mountain tribes), who most likely spoke an Indo-European language, invaded Babylonia and established the Kassite Empire all over Mesopotamia that lasted until about 3,180 years ago.

In northern Mesopotamia, in the time of Hammurabi, another Semitic population, the remnant of an old Akkadian dynasty, established itself as the Assyrian Empire between 4,000 and 3,200 years ago. These people made peace with the Kassites, but they were under constant check by the Hurrians, who eventually toppled their dynasty and ruled over them. The Hurrians are believed to have originated from a region between Lake Van in eastern Turkey and the Zagros Mountains. Their early presence is documented east of the Tigris. Later they spread west, reaching the Levantine shores, and established the Hurrian state of Mitanni in today's Syria around 3,570 years ago. Less than two centuries later the Hurrian state, undergoing a bitter and bloody civil war, was invaded and conquered by the newly revived Hittite kingdom around 3,390 years ago.

THE "SEA PEOPLES"

The Hittites continued their dominion over the Levant and parts of Mesopotamia. They fought and defeated (although the latter is contested by some) the Egyptians at the Battle of Kadesh 3,298 years ago and established their full control over the Levant. The Hittite rule in the eastern Mediterranean was not repressive: Many cultures within the Levant thrived and maintained some level of independence through the establishment of treaties or alliances with the Hittites. But this mode of governance proved fatal for the Hittites, for when the "Sea Peoples" attacked them around 3,220 years ago, their Levantine allies abandoned them, failing to provide the needed support to defend the kingdom.

After the Sea Peoples completely destroyed the Hittite state around 3,205 years ago, they were also able to free the Levant from Egypt's gripping tentacles. These invaders wreaked havoc across the entire Levant and destroyed many of the coastal cities, from Syria to Palestine, where they established their main settlement. While the Hittites were dominating Anatolia and the Levant, the Assyrians, taking advantage of the Hurrian defeat, expanded their own power and encroached on Upper Mesopotamia. They established their capital at Nineveh, on the eastern bank of the Tigris River. They defeated the Kassites around 3,240 years ago and reached and conquered Babylon. A decade later the Assyrian Empire witnessed its first decline. For centuries, the Assyrians had been under assault by the Elamites from the east and the Arameans, who were gaining power throughout the Levant and Mesopotamia.

CHAPTER 12

The Early Tribes

Cultures, like matter, don't disappear—
they morph

BEHIND EVERY EMPIRE WAS A TRIBE

Not until several centuries after the decline of the first Assyrian Empire did the Neo-Assyrian Empire reach its supreme power, around 2,720 years ago. The Neo-Assyrian Empire dominated the area that extended from Mesopotamia to the Levant for five hundred years, after which it was finally invaded by the Medes, who arrived from Iran. In their assault on Nineveh and the Neo-Assyrians, the Medes allied with a Babylonian king from the Chaldean dynasty, Nabopolassar. His son Nebuchadnezzar II brought glory and magnificence back to Babylon. When the Levantine populations were snubbing his dominion, he led a successful military campaign westward and subjugated all the city-states of the Levant, crushing all those who resisted. He captured the last standing Phoenician city, Tyre, and wrecked the Judean capital, Jerusalem, 2,611 years ago, deporting most of its inhabitants to Babylon. He extended his dominion to the outskirts of Egypt, ending several millennia of Egypt's dominance over the Levant.

The Medes were succeeded by the Persians, who, under Cyrus, wasted no time before invading and laying waste to the Chaldean dynasty in Babylonia. Around 2,563 years ago, Babylonia succumbed to Cyrus, and with that, a rich, tumultuous chapter in world history came to an end.

CULTURE OR CIVILIZATION?

Is *culture* interchangeable with *civilization*? A respected archaeologist once told me: "One has to be careful with the use of the term *civilization*." When I asked what word I should use instead, he responded, "Why not *culture*?"

A civilization is the state of a population, delineated by a time bracket and a geographical location. It is a rigid description of a multitude of events that culminate in a set of descriptors and attributes. It is not evolving, and it always ends at a specific time period, usually with destruction. The western Roman civilization lasted more than five hundred years, while the Greek civilization lasted nearly 750 years. Cultures, by contrast, are usually immortalized—they interact with other cultures and breed continuity. They are inherited, and they evolve. The first culture in the world, the Sumerian, blended with the Akkadian and has continued to evolve until this very day. Cultures rarely die, but when they do, an entire history dies with them. Populations with no cultures are like ghosts, without history.

A high-ranking foreign diplomat once invited me to a dinner where many politicians were present—I was the only nonpolitician. The discussion was about extremism and how it had gotten so dangerous, especially in the Levant, where it permeated every community. Everyone was giving an opinion about how to get rid of it or contain it, and their answers involved either wars or some political or economic scheme. Finally, when they asked my opinion, I replied with one word: "Culture." They ignored me and continued their discussion as if I were joking. They, as politicians, were not in the mood for jokes when discussing a serious matter like extremism. I interrupted and said that what I meant was that we need to preserve the many cultures that exist in the eastern Mediterranean, no matter how many there are, and that if we were to do so, the Levant would be a fantastic place, a multicultural oasis. Accepting the other enriches oneself. By preserving cultures, our net gain could be peace not war.

After I finished, I felt good, but they fell silent. I realized how naïve or idealistic I was. It was the first and last time I was invited to such a dinner.

The cultures behind the empires that were described in the previous chapter were essentially tribal and very small. It is not possible to write about them all, but here is a succinct description of some of the early tribes behind the cultures that dominated the Levant in recent millennia.

THE ARAMEANS

As Neo-Assyrian dominance in the Levant and Mesopotamia declined, the Aramean tribes expanded from several local kingdoms in the northern Levant. They established city-states like Sam'al in Cilicia (on present-day Turkey's southern border with Syria) and Beit Agoushi (in Hama, Syria). They extended their reach first to northern Mesopotamia, where they established several little kingdoms, with Guzana (Tell Halaf) at their center. Although the Arameans did not establish a dynasty or a political unit, they expanded as groups and spread farther south into Mesopotamia, reaching Babylon. Aramean power did not last more than a few hundred years, but the Aramean language was simpler to write than other languages of the time. It supplanted Akkadian and the many other languages spoken there and became the lingua franca of the Levant and Mesopotamia. During Persian rule, starting with the Medes around 2,600 years ago, it was used as the political language. Later it was adopted as the main language spoken throughout the entire Persian Empire, which extended from Mesopotamia to the eastern coast of the Mediterranean, reaching Asia Minor in the north and Egypt in the south. Aramaic dominance contracted during the Hellenistic period but expanded again with the Roman conquest that reunified Southwest Asia under one rule. As the language of Jesus Christ, it spread farther and evolved into two branches: Western Aramaic, similar to ancient Aramaic, with several dialects represented by Christian and Judean Aramaic and Eastern Aramaic, noticeably different from ancient Aramaic and spoken in Babylonia. After many centuries, Aramaic gave way to Arabic, another Semitic language that spread at a remarkable speed with the Islamic expansions of the eighth century CE.

Aramaic is still spoken today by the inhabitants of a few

villages in Syria, the most important of which is Maaloula, a mountain town situated about fifty-five kilometers north of Damascus. This town was captured by extremist elements during the Syrian war in 2013 but was fortunately spared the overwhelming destruction seen in other places in Syria.

THE HEBREWS

The Hebrew story begins with biblical Abraham, who departed from the city of Ur in Mesopotamia sometime in the early second millennium BCE to journey to the Land of Canaan. After several decades and an arduous journey that took him through Haran (Turkey), Syria, and Egypt he finally settled in Hebron: "And Abraham left the desirable Sodom and Gomorrah and sojourned at Hebron in Mamre where he heard again from the Lord and built an altar" (Genesis 13:18). After a period of unrest and famine in Palestine, the grandson of Abraham, Jacob, moved and settled in Egypt. Under Jacob's son Joseph, the Hebrews were forced into an exodus from Egypt. Moses led them back toward Canaan but died before they reached Canaan. Joshua, his most faithful disciple, who was born in Egypt and hidden in the Nile as a child to escape being killed during Pharaoh's order to slay all male newborn children, succeeded him in leading the Hebrews. The Hebrews finally made it to Canaan around the second half of the thirteenth century BCE.

When the Sea Peoples were wreaking havoc around 3,220 years ago on the coast of the Levant, burning and abandoning most of its cities, a group called the Philistines appeared and established control over Palestine. It is not known whether the Philistines belonged to the Sea Peoples or were a separate group that came with them, but what is certain is that they arrived in Palestine from an unknown place around the time the Sea Peoples were invading the Levantine coast. They most likely got their name after settling in Palestine. A recent study that investigated ancient DNA samples recovered from Late Bronze Age/Early Iron Age (3,720 to 3,200 years ago) burials in Ashkelon suggests that the Philistines may have come from Crete or other parts of Europe, supporting the theory that the Philistines came with the Sea Peoples.

King Saul united the various Hebrew tribes, who then fought and defeated the Philistines 3,000 years ago. But it was King David who consolidated the Jewish tribes into one undivided kingdom around 2,900 years ago and established Samaria as its capital. Several other tribes that were also occupying the southern Levant, like the Edomites, Moabites, and Ammonites, all united under the rule of the Hebrew kingdom.

Under David's son Solomon, the kingdom prospered and achieved new heights of glory. After Solomon's rule, however, the kingdom was divided into a northern and a southern kingdom and lost power in the face of the expanding Neo-Assyrian Empire. Samaria fell to the Assyrians around 2,740 years ago, then several decades later to Nebuchadnezzar II, the Babylonian who deported a large number of the Hebrew population to Mesopotamia. About fifty years later, the Persians defeated the Babylonians, and Cyrus the

Great, after invading the Levant and capturing Samaria, brought the deported Jews back to their land.

THE CANAANITES AND THE PHOENICIANS

In the northern Levant, the Canaanites were spared the brunt of the destruction wreaked by the Sea Peoples—they thrived and established many independent city-states. In return for being spared, they supposedly provided the Sea Peoples with maritime transport along the Levantine coast that they occupied. The main city-states that made up the Canaanite culture were, from south to north, Accho in Palestine, Tyre, Sidon, Byblos in Lebanon, and Ugarit in southwestern Syria. The Canaanites and the Hebrews adopted a peaceful approach to one another—they were both more interested in commerce and trade than in attaining hegemonic power and making territorial gains, unlike most of the cultures around them.

The term *Canaanite* was used as early as the third millennium BCE, but Canaan as a culture emerged as a strong political entity in the second millennium BCE, when Egypt was loosening its grip on the Levant. Egyptian hegemony over the Levant first faltered with the arrival of the Hyksos. Somewhat mysterious in origin but most likely a Semitic people from the east, the Hyksos established themselves peacefully in lower Egypt, first in the city of Avaris. They somehow managed to gradually dominate almost all of Egypt for one hundred years (3,662 to 3,554 years ago), and they founded the Fifteenth Egyptian Dynasty. The Hyksos developed sophisticated weapons made of bronze and introduced horse-drawn chariots into Egypt, but whether they took Egypt by force remains to be corroborated. Ahmose, the king of Egypt from 3,573 to 3,548 years ago, managed to defeat the Hyksos and recovered Egypt using the very same warfare technology the Hyksos had developed. Ahmose founded the Eighteenth Dynasty. Barely recovering from the Hyksos, Egyptian expansion to the Levant was kept in check by the Hittites of the north. While Egypt experienced this period of crisis and power struggle, the Canaanite city-states prospered, enjoying independence and self-control.

For reasons still unknown, the "Canaanite" label morphed into "Phoenician" during the Iron Age, around 3,220 years ago. The word *Phoinikes* was first used in the Homeric texts to describe the people who occupied the narrow strip of the eastern Mediterranean coast (basically the Canaanites), but Homer's choice of this word to describe them remains a mystery. The most widely accepted explanation is that *Phoinikes* derives from the Greek noun *Phoinix,* which means "purple," in reference to the purple dye extracted from crushed murex shells that was one of the main hallmarks of the Phoenician culture.

The debate over whether the Canaanites were themselves the Phoenicians is still ongoing in the minds of some, but not in mine. The term *Phoenician* is mostly associated with, but is not limited to, the people who occupied the coastal city-states of Tyre, Sidon, and Byblos. While the Canaanites had been occupying the same territory since the third millennium BCE, archaeologists often state that Phoenician autonomy emerged only around 3,220 years ago. This date is absurd, as it coincides with the time when the Sea Peoples were wreaking havoc on the entire Levantine coast.

The Sea Peoples may have entered into a pact with the Phoenicians and spared them a fight in exchange for maritime favors. Most archaeological evidence, however, favors a model of continuity or unity between the Canaanites and the Phoenicians, and many argue that they are one and the same people. Certainly the Phoenician culture emerged from the land of Canaan, and DNA work on the Phoenicians strongly supports the argument that they were the same people who occupied the same territory well before 3,220 years ago (the topic of the next chapter). In the remaining chapters, I will use the words *Phoenician* and *Canaanite* interchangeably to refer to the people who occupied the coastal Levant from Accho to Ugarit, the land of Canaan since the establishment of Byblos as a city around 6,000 years ago.

When Phoenicia emerged as a cultural entity, it was not united under one ruler. Many of its city-states were independent entities, the most powerful of which were the coastal cities of Ugarit, Byblos, Sidon, and Tyre. Byblos was established on a coastal hill where a Neolithic settlement once stood and has been continuously inhabited ever since. It is one of the few cities in the world, if not the only one, that has been continuously occupied for nearly six thousand years. From an early fishing city, it became highly developed in the third millennium BCE, privileged by its favorable geographical location and its natural ports, making it the perfect maritime city of the eastern Mediterranean.

The Phoenicians were explorers—nothing mattered more to them than to explore new territories. From Byblos with its natural sailing ports, they took to the sea and never looked back. They were seeking new territories not only to trade with but to communicate with. In their early exploration days, they would come up to a foreign shore, place some of their merchandise there, go back to their boats at sea, light a fire, and wait. They would wait at sea until they were satisfied with what the locals traded their goods for. They did this at every trading post they established. And almost every trading post became a Phoenician settlement, not a colony.

As the Minoan influence on the Mediterranean was weakening and the Neo-Assyrian Empire's dominion over the Levant was intensifying, the Phoenicians looked to the western Mediterranean for new opportunities. Starting in the early first millennium BCE, they took to the Mediterranean from their coastal cities of Byblos, Tyre, and Sidon. Each of these city-states would send a separate exploratory fleet, and in doing so, they each established trading posts or settlements in almost every island across the Mediterranean. In exploring the Mediterranean, however, they avoided the Greeks, and with only a few exceptions they did not attempt to settle or trade where the Greeks had already established a presence. There is no recorded history of any maritime battle between these two cultures. Their purpose was to explore, trade, and most important interact socially and culturally with their receptive hosts. In establishing their settlements across the Mediterranean basin, they considered themselves not occupiers but migrants seeking to coinhabit and blend with their hosts. They were interested not in territorial gains beyond their well-coveted homeland but rather in cultural exchange. The Phoenicians were not colonizers; wherever they docked their ships, they settled and blended with the host population.

Around the first millennium BCE, the Phoenicians exported cedarwood, which was plentiful in their original homeland, in large quantities to their southern neighbors. It is said that the Temple of Jerusalem during the reign of Solomon was built of cedarwood that Hiram, the king of Tyre, supplied to Solomon in return for other commodities. The Egyptian pharaohs sought Lebanese cedarwood so keenly that they established several trading pacts with the Phoenicians. As they moved farther west across the Mediterranean, the Phoenicians established more trading posts followed by settlements as their commercial activities expanded. They spread their know-how, goods, charm, colorful dyes, cedarwood, and most important their alphabet throughout the Mediterranean and beyond.

The Phoenicians started to settle along the western Mediterranean coast around 3,220 years ago. This was well before the Greeks, who did not start colonizing the western Mediterranean until the eighth century BCE. As early as the second millennium BCE, the Phoenicians made firm contacts with Kition (on the island of Cyprus), which is only 240 kilometers west of Byblos, and established several settlements there. In fact, they established so many settlements across the island that it was hard to distinguish it from the Phoenician mainland. They likely had active trading posts on some of the Greek islands as well, but these never materialized into full settlements. As they moved west, they dotted the North African Mediterranean coast with trading posts and settlements. Some remained relatively small, and others became major settlements, like Leptis Magna in Libya, and Utica and Carthage in Tunisia. As they sailed farther west, they discovered silver on the Iberian Peninsula, where it was not fully exploited by the locals. They traded other goods for it and exported it to Greece and parts of Asia. The silver trade was extremely profitable and was the main reason the Phoenicians established settlements on the Iberian Peninsula. As they expanded across the Peninsula, they also discovered and actively traded tin and copper. Sicily, Sardinia, and Malta became major parts of their trade network and constituted substantial Phoenician settlements for several centuries.

ALEXANDER'S SIEGE OF TYRE

Alexander the Great fought the Persians and defeated them in several key battles as he moved to expand his grip eastward into the Levant. He marched on Phoenicia with his massive army 2,356 years ago and laid siege to Tyre, the motherland of the Carthaginians, and was held at its gates. The Tyrians refused to surrender, and after a siege that lasted six months with heavy losses on the Greek side, Alexander's army finally broke into the fortified city walls, brutally massacred thousands of its inhabitants, and all but destroyed the ancient city.

No one knows the exact reason for this brutal onslaught. Was it because of Tyre's strategic location at the entrance of the eastern Mediterranean, where it provided safe passage to Alexander's army marching south to Egypt? Or did Alexander want to set an example and punish the Tyrians for the fierce resistance they put up against his army? Here are two other possible reasons why he would punish the Tyrians so harshly. First, the Phoenicians had been instrumental in helping the Persian army under Xerxes build a land bridge over the Hellespont (the present-day Dardanelles) during the invasion of Greece. In Alexander's mind, any Phoenician port was a Levantine port of Persia and was to be subjugated. The second possibility: Alexander had a habit of presenting offerings to the Greek gods at temples and other places of worship along the many paths of his numerous conquests. It is said that the Phoenicians of Tyre refused to permit him to do so, which caused the outburst on Tyre.

From Tyre, Alexander went to Egypt and founded the city of Alexandria, then returned to Tyre to consolidate his army in preparation for battle with Darius, the king of the Persians. At the decisive Battle of Gaugamela (Arbela, or modern-day Irbil) around 2,356 years ago in northern Iraq, Alexander defeated the Persians and put an end to their centuries-long rule.

THE FOUNDING OF CARTHAGE

Carthage was by far the most important and powerful settlement the Phoenicians established outside their homeland. It was founded around 2,838 years ago by Elissa, the sister of Pygmalion, the king of Tyre. When Pygmalion was assassinated by his rivals in a power struggle for the Tyrian throne, Elissa was forced to flee, fearing for her life. After a short sojourn in Cyprus, she sailed west with dozens of her followers and docked on the Tunisian shore. The legend goes that she was able, using her Phoenician trading skills, to negotiate with the local leaders and acquire a piece of land in exchange for an ox hide. When the locals agreed that they were willing to exchange a piece of land as big as her ox hide, she cut the hide into narrow strips and laid them in a large circle around a hill, which became the acropolis of Carthage. The acropolis hill was in fact called Byrsa, which in Greek is *bursa,* meaning "ox hide."

Elissa may have had a tragic death soon after she established her settlement (she took her own life after refusing to marry by force one of the local leaders), but Byrsa prospered into a powerful city that dominated not only the North African coast but the Iberian and Italian coasts as well. Compared to the Phoenician mainland and other Mediterranean settlements, Carthage was unique in more than one way; it became the most dominant Phoenician city in the Mediterranean and it had a powerful army, unlike any other Phoenician city. The Carthaginians waged wars and were evidently interested in territorial gains. They fought the Greeks and defeated them in Sicily around 2,574 years ago, then in Sardinia around 2,564 years ago.

Later, while the Greeks were battling the Persians in West Asia, the Carthaginians battled the Romans, initially for control of Messina and later for Sicily and other settlements in the western Mediterranean. The Carthaginians and the Romans fought three wars in total (from 2,289 to 2,170 years ago), referred to as the Punic Wars. The Carthaginians were referred to as "Punic," but why? According to one of the most accepted theories, the Romans, when transcribing the word *Phoinix* into Latin, swapped the first consonant sound (f→p), which gave rise to the word *Punic*.

During the Second Punic War, Hannibal, the Carthaginian general, proved to be a military genius. Too shrewd for the Romans, he dealt them many punitive defeats. His genius lay in the fact that he never fought on his own land. All his battles with the Romans were fought in Europe, across the Alps and the Pyrenees, on Roman soil. He preferred to take the fight to them rather than fight in his homeland. The Punic Wars witnessed movements of armies and people over a wide geographical area in the western Mediterranean and lasted over a century, which made the Carthaginians weary of these wars.

Toward the end of the Second Punic War (2,242 to 2,225 years ago), Publius Scipio, a Roman general, landed in Africa. He had secured alliances with the Numidians, tribes that roamed North Africa, mostly in what is now Algeria. The tribes were divided into the Masaesyli to the west and the Massylii to the east. The Massylii were allies of the Carthaginians but turned on them and sided with Scipio. Their king, Masinissa, with the help of Scipio, defeated the Masaesyli and established a unified Numidian Empire in North Africa that lasted from 2,226 to 2,050 years ago.

Hannibal returned to Carthage to fight Scipio and to defend his homeland, in his first battle on Carthaginian soil. He was decisively defeated in the Battle of Zama (2,226 years ago) and was forced into exile. It is said that he took his own life in an unknown place somewhere in Syria around 2,207 years ago. After his death, the third and last Punic War ended with the complete destruction of Carthage around 2,170 years ago.

Legend has it that the Romans plowed and salted Carthage before they burned it, but this must be taken with a pinch of salt. With the defeat of the Carthaginians in the western Mediterranean and the destruction of Tyre in the east, two of the last standing Phoenician cities, a maritime culture that had dominated and controlled movement across the Mediterranean for many centuries was vanquished.

WHERE DOES GENETICS FIT WITH ALL THIS HISTORY?

The history of the Levant is complex, and the study of the genetics of the successive populations that wove its human fabric is difficult. Most of the cultures described in this chapter were the products of the early populations that moved into the Levant after the Last Glacial Period. Since their genetic signatures have become so amalgamated, it is very difficult to pick them apart. Some, however, have recognizable genetic characteristics that were the result of certain cultural attributes or major historical events that led to their isolation or *en masse* movement from one place to another.

To where did the Phoenicians vanish?

PART VI

THE PHOENICIANS and THEIR ALPHABET

CHAPTER 13

The Phoenicians

Do not believe the Greeks or the Romans; the Phoenicians were not colonizers!

HERODOTUS, THE MASTER STORYTELLER-HISTORIAN, WHO often mixed legends and facts to make history exciting to read, recounted that the Persian king Xerxes ordered his army to build a canal across the Hellespont (today's Dardanelles) when battling the Greeks. Here is how Herodotus described the Phoenicians at work:

> Most of the men engaged in the work made the cutting the same width at the top as it was intended to be at the bottom, with the inevitable result that the sides kept falling in, and so doubled their labor. Indeed, they all made this mistake except the Phoenicians, who in this—as in all practical matters—gave a signal example of their skill. They, in the section allotted to them, took out a trench double the width prescribed for the actual finished canal, and by digging at a slope gradually contracted it as they got further down, until at the bottom their section was the same width as the rest.

THE STIFLING OF A CULTURE

In 2005 I was awarded, along with my colleague, a small grant from the National Geographic Society to conduct a DNA study on the Phoenicians. My colleague and I had met at Harvard while I was doing my postdoctoral fellowship in population genetics. Over dinner one night in a small, crowded restaurant and after a few glasses of good Napa Valley red wine, maybe quite a few, I asked him, What if we investigate the Phoenicians? They were an enigmatic and exotic (for him, a Texan with Irish and Scottish heritage) culture that had several characteristics that would be ideal for a nice DNA study, and more important to me, it was a culture that I relate to most intimately, being Lebanese.

We submitted a short proposal to the National Geographic Society, and they funded us. It was a modest sum but enough to conduct a pilot study, and we were both extremely happy with it.

I was looking forward to meeting with the director general of antiquities at the National Museum of Beirut to tell him about our exciting project and that it was funded by National Geographic. It was all that I could think of on my flight from Boston to Beirut. I met with the director general, and I explained that I was there to collect DNA from ancient Phoenician remains and would like to have his permission. He glanced over at two smirking colleagues who were with us at the meeting, then looked back at me and said sternly, "Why don't you take a tour of our museum and tell me whether you see the name or the label 'Phoenician' on any of the artifacts displayed there." He went on proudly, saying: "Here in Lebanon at the National Museum, we recognize our archaeological artifacts by date, not by cultures."

He certainly caught me off guard, and before I could utter another word, he shrugged and said: "You are certainly one of those Christian nationalists trying to push the theory that the Lebanese are the descendants of the Phoenicians and are not Arabs." I was not sure whether I felt angry or sad, disappointed or betrayed. Perhaps I had been away from Lebanon for too long and lost touch with a new postwar reality that made people less trusting of one another.

When I had discussed the project with my colleague in Boston, I had not thought about religion, or context, and neither of us had thought about the civil war that had ravaged Lebanon for more than two decades. Surrounded by students and academicians, we were two scientists trying to solve an enigma that meant to each of us a different challenge. For my colleague, it was investigating an exotic population; for me, it was investigating my past.

Ignorance was the first thought that came to my head as I left the meeting at the museum. Had I been so naïve and ignorant? Was there nothing left to explore of this enigmatic, lustrous, and exotic culture that had existed in my mind since I was a child?

In fact, what the director general told me was true—when I visited the National Museum of Beirut, I saw that the word *Phoenician* was not present on any of the displayed artifacts. Did they really believe that by ignoring or burying a heritage over some political or philosophical arguments they could make it disappear? I realized then how vital the project I was about to embark on was to me and to my generation.

SEARCHING FOR PHOENICIAN DNA

I grew up in the Phoenician homeland, not far from the only remaining cedar forest that existed during the time of the Phoenicians. I was mesmerized by the stories about their invention of the alphabet, their purple dye, the cedarwood they used in making their beautiful ships, and most important their sea voyages across the Mediterranean. It was my fascination with this culture and my passion for learning about past cultures that compelled me to undertake my first population genetics study. My colleague and I wanted to search for Phoenician genetic traces, or signatures, within modern populations across the Mediterranean.

But a deeper and more gripping and captivating sense of duty or responsibility made me want to channel my research efforts into the Phoenicians. I had left Lebanon, my "Phoenician" homeland, during difficult times. Lebanon, a country that has probably been continuously inhabited for at least 15,000 years, was being torn apart by rivals, each defending a skewed and twisted doctrine. These ill-fated doctrines, based on narrow fundamentals like religion, a fake sense of identity, and pseudo-national ideals, swept not only my homeland but the entire Levant into destruction.

I became unsympathetically sensitive to the term *national identity,* even more so to the pan-national identity that became an accepted doctrine across the Levant. I felt that baseless, fake, and dangerous concepts of pan-nationalism were stripping me and my generation of a true sense of belonging as well as liberty of expression. Belonging to a community that embodies and represents aspirations, ambitions, and ways of being became a necessity for people of my generation. We wanted a community where our culture and heritage could be truly and freely manifested. I could not accept pan-nationalism, nor fake national identities. Pan-nationalism effaces real identities; it buries heritage and has no place for multiculturalism.

The DNA project focused initially on the Y chromosome and, in particular, the region of the Y chromosome that has no homologue to pair with. The first reason I chose this part of the Y chromosome was because it cannot recombine, meaning that every change that is likely to occur on that part of the Y chromosome is caused exclusively by mutations and not by recombination (genetic shuffling through mating/marriage); hence the number or frequency of these mutations is proportional to the time elapsed (number of generations). Basically, it is equivalent to how many times the male gamete (the germ cell carrying the sperm) has divided since that particular mutation (DNA change) occurred, because it is when DNA divides that mutations are transmitted. The second reason was that we presumed it was male Phoenicians who sailed in search of places across the Mediterranean to trade with and eventually settle in, and they carried the Y chromosome.

HISTORY, ARCHAEOLOGY, AND GENES

The main challenge was to figure out a way to distinguish those Phoenician males who left Phoenicia to sail west from the many others who similarly moved from the eastern Mediterranean. Human genetic history can be quite complex; multiple population movement events from different times but with similar migratory patterns or paths are often overlaid. The first and foremost complication was the Neolithic expansion that brought many people, much earlier than the Phoenicians, from the Levant into Europe. The Greeks also expanded west, and the Jewish Diaspora brought people from the Levant, the Phoenicians' neighbors, into Europe over many centuries. These events and perhaps others that went unrecorded can confound and confuse scholars studying the spread of the Phoenicians westward and may mask any Phoenician signatures that were left.

To avoid these confounding factors, we needed to tease out, from all these expansions, the male Phoenician sailors who may have carried their genetic signatures from their homeland in the Levant to the various trading posts or settlements across the Mediterranean. If we could identify unique signatures in the places where the Phoenicians had settled in the Mediterranean, and if these signatures were absent from sites where the Phoenicians did not settle, it would be the best evidence for Phoenician settlement events.

Distinguishing between non-Phoenician and Phoenician-specific patterns was not an easy task. What helped was the fact that the Phoenicians avoided locations where their rivals, the Greeks, were expanding to in the same period (second and first millennium BCE). Historical and archaeological records provided a clear map where Phoenician and Greek settlements or trading posts throughout the Mediterranean were established. Several of these Phoenician and Greek settlements were identified, and DNA was collected from their modern inhabitants for analysis.

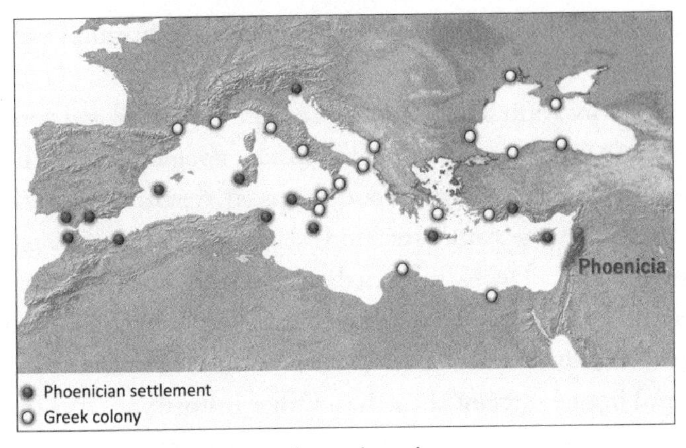

Phoenician and Greek settlements across the Mediterranean (first millennium BCE*)*

Comparing the DNA of two ancient settlements does not provide nearly enough evidence to establish Phoenician-specific signatures. First, one needs a set of comparative pairs of locations, one Greek and one Phoenician, that are geographically close. Second, one must establish common genetic links, or signatures, that exist between all or most of the Phoenician settlements and that do not exist in any of the Greek settlements. Finally, one needs to show that these signatures are anchored in the Phoenician homeland, Lebanon.

The selected Phoenician settlements that we studied were in Cyprus, Malta, Sardinia, Ibiza, and the southern coast of Spain. However, one settlement was key, the ancient city of Carthage, in modern-day Tunisia. Carthage and North Africa were not part of the Neolithic expansion, and based on historical writings and archaeological evidence, it was nearly certain that the Greeks did not settle or colonize that part of North Africa before or during the Phoenician expansions.

Each location selected as Phoenician was paired and compared to other locations, not too distant geographically, that were not settled by the Phoenicians. It would not be surprising that the paired locations shared similar genetic patterns, as they were subjected to similar events. But they would be expected to differ specifically in their Phoenician genetic influence if there had been a transfer of genetic material in one but not the other. Other historical events that were accompanied by population movement would not produce distinct patterns but would leave similar genetic patterns across the landscape.

One could also argue that some of these distinct signatures might be false or might have arisen by chance unless they could be traced back to the Phoenician homeland, Lebanon. After all, Phoenician sailors, merchants, and other males started their voyages from Tyre, Sidon, or Byblos and set sail to islands and coastal lands where they settled. These sailors left sons and relatives behind with similar Y genetic signals that can be easily identified.

THE PHOENICIAN GENETIC SIGNATURES

DNA analyses revealed one genetic signature, the Y haplogroup J2a2, that is consistently preponderant across all the Phoenician settlements tested. But this same haplogroup had a high frequency as well in the Greek settlements tested and hence could not be designated the Phoenician haplogroup. This is not surprising given that eastern Mediterranean populations are genetically very similar, and haplogroup J2 is one of the most common Y haplogroups in these populations. Additional analyses were applied to see whether differences between the Phoenician and Greek J2a2 haplogroups could be discerned. These additional analyses entailed the identification of J2a2 haplotypes. A haplotype is a specific DNA pattern found in a given Y haplogroup, say haplogroup J2a2. These patterns indicate how closely related two J2a2 haplogroups are. Haplotypes can be meaningless when compared among different haplogroups, and they are indicative only when they are compared within one haplogroup. Comparing the haplotypes of two individuals in the J2a2 haplogroup can determine how closely related they are.

A haplotype defines a signature for a given haplogroup. It might be useful to take cars as an example and compare a haplogroup to a car make, say Volvo. Now Volvos come in several colors, and each color would be the haplotype, except that a haplotype is based on more than one indicator (color). In the analyses, eleven indicators were used to determine a haplotype. Each indicator has a relative frequency, some rarer than others. So the more indicators used, the rarer the haplotype, and the rarer the haplotype, the more discriminatory the analysis. In the Volvo analogy, one can use color, type of transmission, engine size, etc. One would also find other car makers that share the same indicators, but these would not be of interest since the focus is on one car make, Volvo (haplogroup J2a2). The chance of finding these indicators present in one kind of car becomes less and less frequent as the number of indicators increases.

The study identified six J2a2 haplotypes with eleven indicators. These haplotypes are referred to as the Phoenician Specific Signals. These six signals were very well represented throughout the Phoenician locations tested but were almost absent from the Greek sites. Using the same analyses, we also identified Greek Colonization Signals that showed a reverse pattern to that of the Phoenician signals.

The six Phoenician Specific Signals were not homogeneously distributed throughout the various settlements. Some were more preponderant than others at specific sites, but they all shared common lineages rooted in the Phoenician heartland, Lebanon. These observations most likely indicate distinct waves of settlement at these sites that originated in Lebanon. For any Phoenician signal to remain detectable after nearly three thousand years means that, whatever admixture occurred in these locations since the Phoenician presence, this signal, although diluted, could not be erased. It must have survived through isolation or genetic drift and resisted admixture and additional east-to-west expansions, such as the Jewish Diaspora, and subsequent population movements to remain detectable until the present day.

Three of the six Phoenician Specific Signals were more frequent in coastal Tunisia, the region where Carthage stood more than 2,000 years ago. These three signatures are extremely rare in all of North Africa and were totally absent from farther-inland Tunisian males. These observations not only provide additional support for the Phoenician expansion narrative but highlight an important interpretation that the Roman destruction of Carthage did not eliminate the Phoenician gene pool there.

While haplogroup J2a2 with its six specific signatures spread through Phoenician male lineages throughout the Mediterranean, this does not by any means suggest that (1) these signatures summarize an entire population, and (2) they are exclusively Phoenician signatures. But they represent a large subset of Phoenician males who left Lebanon more than 2,000 years ago and established their settlements to the west. Phoenicians likely spread other lineages as well that were either washed out by admixture or by other population events that left their signal undetectable by our tests. What was surprising, however, was the magnitude and the spread of the Phoenician signature identified. It has been estimated that this Phoenician signature is currently carried by 6 percent of all the males who were tested across the Mediterranean, which is a very substantial signature.

These signals do exist outside the Phoenician genetic landscape and can clearly be spread by other means. The DNA findings do not suggest that any male with one of the six Phoenician signatures is necessarily a Phoenician! To carry this idea a bit further, a simple genetic test that indicates the presence or absence of a Phoenician Specific Signature does not mean that someone can or should claim Phoenician heritage. DNA is part of one's heritage, and heritage cannot be simplified to a genetic test.

THE DNA OF ANCIENT PHOENICIANS

Y DNA work on the Phoenician settlements across the Mediterranean showed that the early Phoenician male traders or explorers of Tyre and Byblos left a very distinct genetic signature that persisted through time and over many generations. The large percentage of the Phoenician Specific Signals among the modern inhabitants of these early Phoenician settlements argues well in favor of population mixing and integration rather than colonization.

Ancient DNA from skeletal remains in Phoenician burial sites across the Mediterranean indicate that the Phoenician trade networks were also consistent with settlements rather than colonization. Analyses of mitochondrial DNA of ancient samples from Phoenician sites in Ibiza, Sardinia, Sicily, Tunis, and Lebanon show that the Phoenician settlers integrated with the indigenous populations. In the case of Ibiza and Mozia, however, Phoenicians were the first settlers of these islands, and they assimilated European populations from the mainland. Mitochondrial lineages are maternally inherited, since mitochondrial DNA is exclusively transmitted through the mother. In fact, all the ancient DNA work to date shows mitochondrial DNA patterns (maternal inheritance) consistent with female mobility (translocation) across many Phoenician settlements in Europe. Mobility was observed in Phoenician settlements not only in Europe but in Lebanon as well.

Ancient DNA analyses conducted on Phoenician burials at the Monte Sirai site in Sardinia and in Carthage have yielded many examples of unexpected nonindigenous mitochondrial haplotypes. It is now apparent that the Phoenicians may have introduced women to Sardinia and Carthage originating from their homeland (Lebanon) as well as other settlements in North Africa. European mitochondrial lineages have been identified in DNA extracted from skeletal remains discovered in Phoenician tombs in Lebanon. All these observations suggest that female mobility is very well established and must have involved all Phoenician settlements across the Mediterranean, including the Phoenician heartland community.

The ancient DNA work conducted on sixth- to third-century BCE Phoenician burial sites in Sardinia and Ibiza revealed important clues about the spread of the Phoenician culture throughout the West. A high degree of genetic continuity and integration was observed between the Nuragic culture (pre-Phoenician) and the Phoenician culture in Sardinia, findings that are consistent with the archaeological evidence.

This DNA work has brought to life a magnificent piece of my population heritage that had been buried or forgotten. The research findings provide strong foundations that complement the archaeological findings and the written stories about a culture that has not yet been given proper appreciation. To my many friends who ask me whether they are Phoenicians, I say, Do not search for your identity—you must make it yourself, and whatever heritage you have, be proud of it. It is through knowledge of and proper tributes to our ancestors that we can forge and strengthen our modern identity. It is a challenging but a wonderful undertaking to be able to define your own identity.

CHAPTER 14

The First Alphabet

Dating inscriptions on stone or clay tablets is not an exact science

THE ORIGIN OF THE FIRST linear alphabet—who may have invented it and where—is the subject of a continuous debate. The stone and clay tablets on which early inscriptions are found are hard to date with confidence. The anachronistic discoveries of these tablets make the clues they offer about the evolution from pictographic to linear writing very challenging to decipher. Despite the dating uncertainties, however, and the many recent discoveries of archaic linear inscriptions, the proto-Canaanite script, the one invented around the beginning of the second millennium BCE, remains the most likely candidate for the first linear alphabet.

> The supreme specimen of the highest form of human genius that has yet to be improved upon . . . the most definitive of human creations . . . The Phoenician alphabet remains to this day what its first creators intended it to be. And today, four thousand years later, in a world relentlessly challenging and questioning all

assumptions, the Phoenician alphabet endures and remains, superseding all the other breakthroughs of human civilization.

Before the Canaanite alphabet was invented around the seventeenth century BCE, at least three forms of writing existed in ancient Southwest Asia: one in Mesopotamia, one in Egypt, and one in Crete. The Egyptian script, called hieroglyphic, was based on pictographic signs (images of objects, animals, and humans). Many hundreds of these existed, and some had ideographic or phonetic values. The Mesopotamian script was also pictographic, with hundreds of signs, but it was simplified by using wedge-shaped lines instead of full drawings, on soft clay tablets, which were baked later to make them hard and lasting. This script, with syllabic value, was invented by the Sumerians and was referred to as cuneiform script, after the Latin word *cuneus,* meaning "wedge." The Cretan (archaic) script was a phonetic script using hundreds of signs, each representing a syllabic value. The Linear A script diverged from the Cretan script, and the Minoans (ancient Cretans) used both for several centuries until they were replaced by the Linear B script around the fourteenth century BCE during the Mycenaean rule of Crete. While the Linear A script was written in a Semitic language, Linear B was based on an Indo-European dialect.

The Levant, wedged between the two dominant powers in the region, the Egyptians and the Mesopotamians (Akkadians in the third millennium BCE, Babylonians in the second millennium, and Assyrians in the first millennium), was obviously exposed to both the hieroglyphic and the cuneiform script systems. Through its maritime connection to the West, it also had significant and continuous contact with the Minoans in Crete.

Archaeological evidence points to Byblos as a major recipient of every aspect of Egyptian culture, including its system of writing, since the third millennium BCE. In Byblos, during the Egyptian Middle Kingdom (4,064 to 3,806 years ago), a new script was invented. Possibly influenced by the Egyptian hieroglyphic script, the new script was called the Pseudo-Hieroglyph script of Byblos. It was first identified by Maurice Dunand (the main French archaeologist who excavated Byblos in the 1940s), and according to him, it was written in a Semitic language. The Pseudo-Hieroglyph script also shows some similarities with the Cretan pictographic script. Scholars suggest that the Pseudo-Hieroglyph system may have been the mediator of the Cretan Semitic pictographic script, as Crete at that time was heavily influenced by the Levantine maritime culture.

THE PROTO-SINAITIC SCRIPT

In the early twentieth century, a few dozen alphabetic inscriptions were found in Serabit El-Khadem, a mining site for turquoise, in the Sinai, in northern Egypt. These inscriptions were labeled "Proto-Sinaitic" and were found along with artifacts that were dated to around the nineteenth century BCE. The Egyptians have mined this site since the Twelfth Dynasty (4,015 to 3,826 years ago) using miners from the Levant. While scholars debate the dating of these inscriptions, they agree that the script is alphabetic and that it resembles the Pseudo-Hieroglyph script of Byblos.

The fact that the Serabit El-Khadem inscriptions were based on a direct translation of hieroglyphic signs led some linguists and a few archaeologists to argue that this was the first linear alphabet, born in a place that was directly under the rule of Egyptian culture, where hieroglyphic signs were in use. Based on these inscriptions only, they advanced the theory that the first alphabet was born in the eastern delta of the Nile, where hieroglyphics were mostly used. In 2010 the Egyptologist Orly Goldwasser of Hebrew University of Jerusalem argued that the inventors of the first alphabet were illiterate Canaanite settlers from the Phoenician coast who arrived in Tell El Daba to work at the Serabit El-Khadem turquoise mines. Her claim was contested by many, and while these inscriptions may show the oldest alphabetic script found to date, there is not enough evidence to fully support her claim.

In his rebuttal to her publication, Anson Rainey of Tel Aviv University wrote: "The alphabet was invented by highly sophisticated Northwest Semites who knew not only hieroglyphics but probably also hieratic, the cursive script generally used by Egyptians at that time." He went on to argue that the Serabit El-Khadem inscriptions survived because they are etched on hard surfaces. The alphabet was meant to be written on papyrus and not on stone, and we have yet to find any preserved remains. He concluded: "It is obvious that the original pictorial form of the alphabet must have been written on dozens, hundreds, of papyrus sheets that have not survived. The miners who inscribed their thoughts on the walls of the turquoise mines or on the cliff above the smelting camp at Bir Nasib, were hardly the inventors of the alphabet."

In a separate reply to Goldwasser, Christopher Rollston of George Washington University argued that the inventors of the alphabet were Northwest Semitic speakers who were well versed in hieroglyphics.

To continue the argument against the alphabet's invention in Serabit El-Khadem, the Proto-Sinaitic script did not evolve or prosper in situ (Sinai), and it appears that its diffusion was halted when mining at Serabit El-Khadem stopped, coinciding also with the end of the West Semitic–speaking Hyksos then ruling in Egypt. Finally, it would be highly unusual if not puzzling for the first alphabet to originate in a place where it did not prosper or evolve. Consequently, Serabit El-Khadem is highly unlikely to be the original birthplace of the alphabet.

THE UGARITIC SCRIPT

Since 1929, numerous clay tablets with cuneiform script have been discovered in the ancient city of Ugarit (today Ras Shamra), situated on the northwestern coast of Syria. Their date has been bracketed between 3,424 and 3,214 years ago, and their texts reveal a thirty-sign cuneiform alphabet and include poetry about religion, rituals, medicine, law, business, and teaching. Since the discovery of this script, and despite its resemblance to the Proto-Canaanite script, being essentially based on pictographs, many scholars suggested that this cuneiform script may have been independently derived and is not at all related to the Proto-Canaanite script. Some scholars even suggested that while those who invented the Ugaritic cuneiform script had some knowledge of the Proto-Canaanite script, they did not imitate it, and the resemblance is only coincidental.

The phonetic resemblance between the Ugaritic cuneiform script and the Proto-Canaanite script is utterly striking. The corresponding sounds and the phonetic values for each of their respective signs are almost identical. In addition, if the five consonantal signs that are not used in the subsequently derived Phoenician alphabet are removed from the thirty Ugaritic signs, the order of the signs in the two scripts is identical. One can only conclude that the twenty-two-sign Phoenician alphabet and the thirty-sign Ugaritic cuneiform script originated from one common writing system: the Proto-Canaanite script. Three facts are to be retained about the Ugaritic cuneiform script: (1) It dates back to about 3,320 years ago; (2) it is based on the Proto-Canaanite script that existed before 3,320 years ago in the Levant; and (3) the order of the letters (signs) in the Proto-Canaanite script was fixed sometime before 3,320 years ago.

THE CANAANITE ALPHABET

The Canaanites were familiar with the Egyptian hieroglyphic. The scribes of Byblos started to experiment with a linear alphabet in the third millennium BCE. The derivatization of the linear alphabet from the Egyptian hieroglyphic was by no means a quick, simple, or straightforward process. On the contrary, it was complex and slow. The phonetic values in each system were different and evolved widely apart, accumulating significant distinctions (especially Canaanite). Acrophony is a concept that resulted from the idea of replacing hieroglyphic signs with the first consonant of whatever that sign meant in the Semitic language. The scribes of Byblos used the concept of acrophony to develop the Canaanite alphabet, also referred to as Proto-Canaanite.

The Canaanites used simplified Egyptian signs with Canaanite values. For example, the Canaanite phonetic sign for "house" is based on the Egyptian sign and is pronounced in Canaanite as *bet*. Its derived syllabic value when using the linear alphabet was reduced to the first consonant, *b*. In an-

Hieroglyph	Proto-Canaanite	Ugarit	Phoenician	Greek	Latin	Semitic name
				A	A	Aleph
				B	B	Bet
				T	T	Tet
				M	M	May
				P	R	Rosh

The evolution of the alphabet, from hieroglyphic signs to Latin letters

other example, the phonetic sound for the sign of the human head in Canaanite is *ras* or *rosh*. Its derived linear value was reduced to the first consonant, *r*, and so on.

IF IT LOOKS LIKE A DUCK, SWIMS LIKE A DUCK, AND QUACKS LIKE A DUCK . . .

The Proto-Canaanite script was invented in Byblos before the seventeenth century BCE. Some archaeologists, linguists, and historians may take issue with the date that I am proposing. Dating inscriptions, however, is a very difficult and delicate task, as one cannot carbon-date a stone slab and must rely on more indirect means of dating, like dating artifacts or biological remains found along with the inscribed tablets.

Here is why I believe that the first alphabet was indeed the Canaanite alphabet, and that it was invented in Byblos before the seventeenth century BCE. First, several Proto-Sinaitic inscriptions have been discovered throughout the Levant. All those that have been dated prior to the thirteenth century BCE were poorly preserved (mostly written on papyrus) and found either on isolated fragments, like arrowheads and pieces of pottery without an archaeological context with defined stratigraphy, or on stone slabs, which make their dating imprecise.

Most of the well-preserved Proto-Sinaitic inscriptions that were discovered within defined stratigraphic layers were dated not earlier than the thirteenth century BCE. Therefore, these inscriptions were likely derived from the Proto-Canaanite script of Byblos. They were etched by local southern Levantine scribes who were exposed to the Proto-Canaanite script through their documented interaction with the Canaanites or by some Canaanites visiting the southern Levant.

Second, the Ugaritic script was influenced by the hieroglyphic script as well as by the Akkadian (cuneiform) system. The order of the Ugaritic script corresponds almost exactly to the Canaanite alphabet (or Proto-Canaanite script). This confirms that the Ugaritic script can only be derived from the Proto-Canaanite script, in which the order of the signs had already been fixed and stabilized prior to 3,320 years ago. This same principle can be applied to the South (old) Arabic script, which was derived from the same Proto-Canaanite script sometime between the fourteenth and thirteenth centuries BCE.

Third, soon after the discovery of the Linear Cretan script, some advanced the theory that this script was the precursor of the Proto-Canaanite script. This theory has now been refuted, since scripts similar to the Cretan ones were also found in Sardinia, and all the evidence shows that the Cretan script and other archaic Western scripts were in fact borrowed from the seafaring Canaanites of Byblos.

Finally, the linear alphabet could have originated only in a place that had abundant exposure to the hieroglyphic script but was not dominated by it, a place where other scripts were also used and were allowed to be used, and that ideal place was Phoenicia. Trade between Egypt and the Levantine coast had been ongoing since the third millennium BCE. The Proto-Canaanite script was introduced sometime around the thirteenth century BCE and reduced from twenty-seven signs to twenty-two linear forms by the scribes of Byblos. It was established as a fully recognized linear alphabet in the eleventh century BCE.

After that it became widely used in the Levant and was referred to as the Phoenician alphabet. This linear alphabet could not have been invented anywhere else since its evolution is well documented in situ, in the place of its birth, Byblos. If it were invented elsewhere, one would expect it to evolve and prosper in situ. It did not evolve or prosper in Ugarit, and it certainly did not evolve or prosper in the Sinai or the southern Levant.

Further, Byblos is central to Ugarit in the north, the Sinai in the south, and Crete in the west. It was the "Byblians" who had the first maritime contacts with the Cretans after the middle of the second millennium. Byblos and Egypt had a documented cultural exchange since the third millennium. It makes perfect sense that it was the Canaanites of Byblos who radiated their linear alphabet to all these places. Further, almost the entire linear Phoenician alphabet (twenty of twenty-two signs) can be clearly and unambiguously identified on the sarcophagus of the king of Byblos, Ahiram. The sarcophagus dates to the thirteenth century BCE, but its inscriptions most likely date to the tenth century BCE. To date, no prior inscriptions using a complete set of signs of a linear alphabet have been found. This and the fact that the Phoenician alphabet is a more simplified script and less structured in shape than all other preexisting scripts are testimony to its in situ evolution. It was the nature of the Phoenicians' culture and their love of trade that led them to simplify and radiate their linear script. The alphabet was and is indeed Byblian!

THE GREEK ALPHABET

It was the Phoenicians who introduced the twenty-two-character alphabet to the Greeks around the tenth or ninth century BCE and, through the Greeks, to the world in the ninth and eighth centuries BCE.

How and exactly when the Phoenician alphabet was introduced to Greece is not yet accurately documented, but the name Cadmus has always been associated with it. Herodotus writes that Cadmus the Phoenician settled in Boeotia (in central Greece) with his followers and taught their alphabet to the Greeks: "They transmitted much lore to the Hellenes, and in particular, taught them the alphabet which, I believe, the Hellenes did not have previously, but which was originally used by all Phoenicians." Cadmus was also a legendary character, and Herodotus has never shied away from mythology in recounting history. The legendary Cadmus was the son of Agenor, the Phoenician king of Tyre. His beautiful sister Ourouba, or Europa, was abducted, while walking on the beach in Tyre with her friends, by Zeus, who had transformed himself into a beautiful large white bull. Cadmus went after them to Crete to bring her back. The myth further suggests that Europe was named after Ourouba.

Leaving mythology aside, the fact remains that the twenty-two-character Phoenician alphabet was carried to Greece by some Phoenician sailors, and perhaps one of them was called Cadmus. The first tablets with Phoenician writing on them were found in Crete. The Phoenicians of Byblos had contact with the Greeks. They had interacted with Crete since the second millennium BCE. It was believed that the Phoenicians transmitted their linear alphabet to the Greeks around the ninth century BCE, a time when the Phoenicians were also speaking Greek in addition to their Semitic language. However, certain scholars, citing epigraphic comparisons using the archaic Cretan inscriptions, advance the theory that the Greeks may have adopted the Canaanite alphabet around 1100 BCE. Some even suggest that it was the Philistines who adopted the Canaanite alphabet after arriving in the Levant and in turn transmitted it to the Greeks.

Whether it was in the twelfth, eleventh, tenth, or ninth century BCE, the Greeks, after adopting the Phoenician alphabet, made several changes to it by introducing some vowels to accommodate certain sounds used by Indo-European but not by Semitic speakers, and they deleted or transformed some consonants used in typical laryngeal Semitic sounds. For example, the Phoenician *aleph* and *ayn* became, for phonetic reasons, the *alpha* and *omicron* in Greek. The character for the Phoenician sign *saad* was deleted, and the *phi, chi, psi, eta,* and *omega* were added. In total, the Greeks added seven new characters and removed two, making the twenty-two-character alphabet into twenty-seven characters. The Ionian alphabet, which was developed further, gave rise to the modern twenty-six-letter English alphabet I am using to write these words.

The impact of the Phoenician alphabet has not been fully acknowledged, primarily because it was the Greeks who disseminated it throughout the rest of the world, taking advantage of their rapidly expanding empire, while that of the Phoenicians was declining.

To date, centuries of Phoenician cultural evidence remains buried under the concrete of the highly urbanized coastal cities of Lebanon. Most of what we know about the Phoenicians comes to us packaged in Greek or Roman writings. Instances of the Phoenician script that have been found to date are rare and scattered and lie dormant in basements of museums around the Mediterranean. Many more are yet to be discovered, judging by the frequency of the Phoenician artifacts that private landowners unearth while excavating to build houses along the Lebanese coast.

Ancient cultures must not be forgotten, and they should not be described only by those who were their rivals, as was the case with the Phoenicians. It is unfortunate that for the last fifty years, no serious archaeological studies about the Phoenicians have been conducted in Lebanon. Since its inception as an independent country in 1943 and even before, Lebanon has been undergoing a fundamental identity crisis that has left it unstable, reeling from one crisis to another. This identity crisis has consumed most of my adult life and that of my friends. Witnessing it firsthand is what has compelled me to undertake my DNA work on the Phoenicians and to share my findings in this book.

PART VII

RELIGION and the MAKEUP of the MODERN LEVANT

CHAPTER 15

A Complex Narrative

A human journey neglecting the past is like a life devoid of memories

A GENETIC PUZZLE OR A FRACTURED MOSAIC?

The Levant is a peculiar term, one that is loosely defined. Europeans coined it in the thirteenth century to designate some of the eastern Mediterranean countries. The term literally means "the rising," maybe in reference to the rising sun or perhaps the sandstorms or winds rising from the east.

In a broader context, the Levant consists of the coastal eastern Mediterranean running from the mountains of Cilicia in southern Turkey to the northern Sinai Peninsula. The Levant has played a crucial role in human prehistory and history, primarily owing to its strategic geographical location connecting three continents: Europe, Asia, and Africa. It is considered the initial destination for migration waves of modern humans coming from Africa to seek more temperate climatic conditions prior to and following the Last Glacial Period. Furthermore, it served as a corridor for Neolithic migrations from the southern Fertile Crescent to Europe and North Africa.

The Levant is also an abstract designation that refers to a moment in history and not a physical location. It has always been a melting pot, an opulent cultural mix, and a rich crossroads. It is the heart of "Eastern Mediterranea." Most important, it was home to the first out-of-Africa migrants. It was in the Levant that *Homo sapiens* mated with Neanderthals more than 25,000 years ago.

Byblos, Jericho, and Damascus, three of the oldest cities in the world, were founded in the Levant. It is the birthplace of the three Abrahamic religions, both a blessing and a curse. It was a central node for the spice and silk trade between East and West.

COMPLEX GENETICS

The Levant's genetic complexity can be demonstrated in several instances. For example, the phylogeographical distribution of modern Jews in the Levant is subject to many considerations. Assyrians in the eighth century BCE, then Babylonians in the sixth century BCE, deported Jews *en masse* to Mesopotamia; then a few decades later the Persian king Cyrus the Great brought them back to the Levant. At the beginning of the first millennium CE, following the Roman persecution, Jews were forced to disperse again.

Jews have a unique phylogeographical status within the Levant for several compelling reasons. Beginning with the Assyrians around 2,760 years ago and up until very recent history, Jews were forced to disperse from the Levant on multiple occasions. These dispersals led to a Jewish Diaspora that fragmented them, in time and space, into groups, each one characterized by distinct and extensively researched genetic signatures. Over time these groups incorporated notable levels of genetic admixture within their geographical locations. Sephardic Jews genetically form a tight cluster with Levantine groups, while Ashkenazi Jews exhibit closer genetic ties to the Caucasus and Eastern Europe, reflecting historical admixtures with Europeans.

The current Jewish populations in the Levant are largely derived from a complex resettlement pattern originating from multiple sources within the last eighty years, possibly deviating from the pre-Diaspora distribution. Jewish genetics have been subject to more extensive study than those of any other group in the world.

While the post–Last Glacial Period migrations were restricted to a handful of populations that managed to survive, their genetic impact on the modern Levantine populations has been extremely significant, as they constituted almost the entire genetic stock of the modern Levant. In the last several millennia, however, massive population movements of Babylonians, Assyrians, Romans, Crusaders, Ottomans, and more undoubtedly left some sort of a genetic footprint there.

Modern Levantine populations, for example, show varying levels of admixture with their southern neighbors, including Arabians, suggesting that genetic exchange must have occurred. Some of these exchanges must have taken place recently, most likely during the Islamic expansion of the eighth century, which brought people from Arabia into the Levant.

Although recent migrations have undoubtedly made genetic contributions, their impact alone is not substantial enough to account for the observed genetic diversity and its patterns.

EXPLAINING THE CONTRAST

When we are observing maternal versus paternal lineages, we find a sharp genetic contrast between the Levant and the rest of Southwest Asia. With maternal lineages, the affinity is between the Levant and Europe, but with paternal lineages, the affinity is between the Levant and Yemen, Saudi Arabia, and Africa, suggesting the two sexes had different mobility patterns. Movement in both directions between Europe and the Levant most likely involved men and women, consistent with a "demic diffusion" expansion model whereby people expand slowly from their habitats, gaining additional territory. Movement from the Levant into Arabia or Africa, however, was more male dominant and may have involved demanding labor, military, or other activities that made it harder for women to accompany their spouses.

People moved for four main reasons; trade, colonization, military occupation, and immigration. These movements can have a major impact on the recipient population, and each type may differ with regard to gender. Hence it is vital to consider historical, archaeological, and human mobility data when we are using DNA to learn about the ancestry of a population, especially in a region like the Levant.

The differential genetic affinities observed in the Levant have multiple explanations. Since Neolithic times, the Levant has been heavily impacted by Anatolians, who have also contributed significantly to modern European populations. Since Neolithic times, population movement out of Anatolia likely happened through a demic diffusion process, as mentioned earlier. With such diffusion, the mitochondrial genetic affinity between the Levant, Anatolia, and Europe is not surprising, as they together formed a common moving entity during the agricultural expansion; female migration during these times was a lot less favorable and more restrictive (childcare) than male. Furthermore, there is no evidence to date suggesting a significant interruption of the movement of people between the Levant and Europe, but quite the opposite. It has been rather continuous, through commerce and invasion as well as occupation.

THE WARS THAT SHAPED THE MODERN LEVANT

The historical events, mostly wars, that shaped the Levant certainly cannot be summarized in a few pages. But some key events may have contributed to substantial population movements that had a subsequent impact on the genetic pool of the Levant.

THE SELEUCIDS AND THE PERSIANS

For centuries, the Levant was a main battleground between the Greeks and the Achaemenid Persians. These two empires fought until Alexander of Macedon (the Great) defeated the Persian king Darius III 2,355 years ago, ending Achaemenid rule. After Alexander died, his general Seleucus assumed power over Asia Minor, parts of Central Asia, Persia, and the Levant. Seleucus, in addition to extending his control over much of the eastern Mediterranean, built the Seleucid Empire, which lasted several centuries (2,336 to 2,087 years ago). During his reign, to solidify his grip on the Levant, he encouraged the settlement of Greeks in the Levant.

During the Seleucid Empire, the Hellenistic culture dominated the Levant and most of Mesopotamia. Around 2,325 years ago Seleucus founded Antioch (Antakya in modern Turkey) by settling the inhabitants of Antigonia (a Hellenistic city in ancient Syria founded by Antigonus), who were Greco-Macedonians, along with other Greeks and autochthonous (indigenous) non-Greeks, farther down the Orontes River. (Seleucus was himself Macedonian.) He then planted more Greco-Macedonian colonists and expanded the city. The Greek population grew substantially in several Levantine locations. The Seleucid reign privileged the Greek language in the Levant, and many cities and towns adopted Greek names, among them the second-largest city in Lebanon, Tripoli ("triple city" in Greek). The Seleucid Empire was a major colonizing force. Seleucus was eager to sustain his dominion over his eastern land gains; he established many military colonies and brought many Greeks from the overly populated Macedonian region to areas in the Levant when his influence was waning.

Colonization of the Levant continued with Seleucus's heir and son, Antiochus. During Antiochus's reign, colonists came from cities across Greece but also from several parts of Asia Minor and the Aegean. It is not certain how much these settlers contributed to the Levantine genetic pool. Some of them could have been soldiers who stayed only during war periods or who were not necessarily Greek but fought with Seleucus and his followers, a practice common since the time of Alexander. Others claimed Greco-Macedonian origin to gain privileges, but they would have been few in number. What is certain, however, is that Greece was a major contributor to the Levantine populations during Seleucid rule.

My grandmother used to tell me a story, which she in turn had heard from her grandmother, about these ancient Greeks, armed to the teeth, who settled in our hometown a long time ago. She would point to two huge rocks that stand like towers and say: This is the gate to their temple; they called it the Gate of the Wind. Indeed, the two rocks stand at least twenty meters high, like two towers, atop the mountain facing my hometown. They are about ten meters apart, and they do indeed look like a gate.

The Parthians, a group of northeastern Persian tribal kingdoms, rebelled against the Seleucids and ended their rule over Persia around 2,270 years ago. The Parthians became the second Persian dynasty, and their rule lasted five hundred years. They fought the Romans, who had replaced the Greeks in ruling Asia Minor and the Levant, and achieved early victories. But lengthy wars with the Romans exhausted the Parthians, who were also riven by internal struggles, and they were ousted from power by the Sassanids in 224 CE.

The Sassanids, another Persian dynasty, established the Sassanid Empire. They continued their wars against the Romans and regained most of the territories lost during previous centuries after defeating the Roman army and taking Emperor Valerianus prisoner in 260 CE.

THE ROMANS

Back in the Levant, after the Greeks came Pompey Magnus's Romans, who conquered Antioch around 2,088 years ago and later annexed the entire Levant as a Roman province. It was considered one of the empire's most strategic provinces, providing access to the eastern Mediterranean shores with its many ports but also access to the East through well-established trade routes dating back many millennia. The Berytian (modern-day Beirut) school of law, with its native jurists like Papinian and Ulpian, was the most prominent law school of the Eastern Empire.

Today, scattered across the remaining cedar and pine mini forests of the Lebanese mountains, the visitor may see large rocks bearing the carved Roman inscription IMP HAD AUG (Imperial Hadrian Augustus). The Roman emperor Hadrian (76–138 CE) personally ordered these inscriptions in his effort to claim the valuable cedars as his own. Unfortunately, fewer than four hundred ancient cedar trees, occupying an area less than 5 percent of their ancient habitat, remain standing today from what was once a magnificent cedar forest covering the entire western side of Mount Lebanon. It is that same cedar forest that was central to the *Epic of Gilgamesh* (the first literary epic ever recorded, written in Akkadian). Gilgamesh—the king of Uruk—and his friend Enkidu traveled to this mystical forest, where they defeated its guardian Humbaba and cut down its trees, angering the gods. Since the time of Hadrian, many of these trees have been cut for wood and used for heating or construction. The Ottomans during their rule over Lebanon destroyed vast numbers of these trees, and

British troops during the Second World War cut what was left to build a railroad between Tripoli and Haifa.

While the Romans dominated the Levant for many centuries, their genetic impact on the region remained limited. They governed their provinces with "propraetorian legates," or imperial representatives, who relied on the disciplined Roman army but allotted enough autonomy to local leaders to control much of the province. In the Eastern Empire especially, very few Roman administrators were present. Most of the Roman soldiers in the Levant were probably Levantine themselves, recruited to serve in the Roman army. The Levant remained a Roman province until the empire split into the Western and Eastern empires in 395 CE.

THE BYZANTINES AND THE CALIPHATES

In the year 476 CE, the Levant passed from the grip of the Eastern Empire to the Byzantine Empire. During the Byzantine era, the Levant, Asia Minor, and Greece witnessed continuous human mobility primarily due to the many wars between the Byzantines and their rivals but also due to religious pilgrimages. While the genetic record shows that the Romans did not do much interbreeding with local populations, the Byzantines established many permanent movements primarily from west to east.

When the Byzantine Empire replaced the Eastern Roman Empire, it warred against the Sassanids, who were constantly harassed by the Huns and later the Armenians on multiple fronts. After nearly a century of rivalry, the two empires exhausted each other, and neither was ready for the onslaught of the Arabs. The Umayyad caliphate, of the first Muslim dynasty, claimed Damascus as its capital in 661 CE. The Sassanid Empire, drained from its long wars, with a failing economy and multiple internal and external struggles, quickly succumbed to the Arab invaders and crumbled. The Byzantine Empire, although it did not fall, was heavily battered, losing the Levant and Egypt to the Arabs.

Later, the Byzantines regained some of the lost territory, including parts of the Levant. In 750 CE, the Abbasid Arabs brought down the Umayyad dynasty and declared Baghdad their capital.

THE SELJUKS AND THE FATIMIDS

Fighting continued until the beginning of the second millennium CE, when a new dominant power appeared, the Seljuk dynasty. The Seljuks were Oğuz Turkic tribesmen who had dwelled in Central Asia between the Black Sea and the Caspian Sea. The Seljuks waged wars on the Byzantines and the Abbasids, and by 1055 CE they controlled Baghdad, most of the Levant, and Turkey. Around the same time in Egypt, the Fatimids rose against the decaying Abbasid dynasty, already swaying from Byzantine and Seljuk pressures, and consolidated their power over the northern Levant, including the Holy City of Jerusalem.

THE CRUSADERS

When I was growing up in northern Lebanon near the Mediterranean, every morning my ride to school took me past a huge stone citadel perched on a hill on the western bank of the Kadisha (Abu Ali) River, on the outskirts of Tripoli. This citadel stood tall like a huge fortress, dwarfing every other building in the city. This was the citadel of Raymond de Saint-Gilles, or Raymond I of Tripoli. Although it came to be named after him, it was constructed well before his arrival in 1103, by Sufyan Al-Azdi, an Arab commander under Mu'awiyah, the first caliph of the Umayyad Caliphate. Saint-Gilles took it as his headquarters for its strategic location surrounded by a flowing river and overlooking Tripoli, then expanded it into a large Crusaders' fortress.

Overwhelmed by the Seljuk Turkmen and the Fatimids, the Byzantine emperor Alexius I Comnenus called on Pope Urban II to help in his fights against the Seljuks and to regain the lost Holy City of Jerusalem. The pope heeded the call, and in 1095 he made a public appeal for Christians to rescue the Holy City and help the Byzantines in their war against the Muslims. European Christians responded and mounted a dozen military campaigns (Crusades) against the Seljuk and Muslim occupiers of the Holy Land. The four Crusades that had a major impact in the Levant were highly documented, and the men who took part in them came from France, Germany, and England by land and later from Italy by sea. The Crusaders brought significant gene flow from Europe into the Levant over the period of their occupation, which spanned more than two hundred years. After they recaptured the Holy City from the Fatimids in July 1099, they remained in the Levant to protect their captured territory from the Fatimids and the Seljuks.

SALADIN AND THE AYYUBIDS

In the meantime, Saladin (Salah Al-Din), a Kurdish military leader during the Fatimid dynasty, rose in stature after some key victories in Egypt and established his own Ayyubid Sultanate in 1174 CE. In 1187 he launched a major military campaign against the Crusaders. At the Battle of Hattin, a village a few kilometers east of Lake Tiberias, he inflicted a severe blow on their army and retook the city from them. During their third campaign, the Crusaders defeated Saladin in the Battle of Arsuf in 1191, but they never recaptured the Holy City. Instead, they signed a peace treaty with Saladin. In their fourth campaign, the Crusaders captured Constantinople and wreaked havoc on the city, nearly destroying it after a failed attempt to submit the Byzantine church to Rome. While the Crusaders were struggling to save their "kingdom" in the Levant, two major ruthless and merciless forces were marching on the Levant: from the Mongolian plains in the East, the Mongols of the great warrior Genghis Khan, and from Egypt, the Mamluks led by Sultan Qutuz.

THE MAMLUKS AND THE MONGOLS

The Mamluks were Turkic people from the Caucasus and other regions north of the Black Sea whose poor families had sold them into slavery to the Abbasid rulers. They converted to Islam and were skillfully trained in the arts of war and horsemanship and became superb warriors. The Mamluks, led by Qutuz, took control over Egypt around 1250 CE from the Ayyubid Sultanate that had been founded by Saladin.

In 1258 the Mongols, led by Hulegu Khan, nearly destroyed the Seljuk Empire, breaking it into smaller emirates. After ransacking Baghdad, they took over Damascus and much of the Levant. As he was continuing their march to Egypt, Hulegu had to abandon his troops to go back to Mongolia, leaving the command of his army to his lieutenant, Kit Buqa.

Qutuz mounted an army and met the Mongols in Ain Jalut, north of the Holy City, and dealt them a significant blow, nearly decimating their force. This decisive battle led to the retreat of the Mongols from the Levant in 1260.

After their successes against the Mongols, the Mamluks turned their attention to the Crusaders, who had stayed on the sidelines for the most part. The Crusaders lost their last fortress, in the Syrian city of Accho, to the Mamluk army in 1291 CE, and their military presence in the Levant disappeared. The Crusaders' reign was rife with ambiguity, contradictions, and bloodshed. It marked the Levant with hatred that still reverberates.

THE OTTOMANS

The Mamluks controlled the Levant until the Ottomans defeated them in 1517 CE. The Ottomans were one of the emirates that the Mongols had carved out of the Seljuk Empire. They captured Constantinople, the capital of the Byzantine Empire, and claimed it as their own in 1453 CE; Constantinople officially became Istanbul in 1930. The Ottomans dominated the Levant from 1517 CE until the end of the First World War, when they were defeated by the Allies.

The above events were associated with massive population movements that undoubtedly had an effect on the genetic makeup of the Levant. Some of them, like the Roman and Ottoman occupations, did not have a major genetic impact, as can be deduced by the governance style of the occupiers. The genetic work thus far has supported this view. Other events most likely had sizable genetic footprints: The Crusades, the Seleucid reign, and the Islamic expansion all saw massive population displacement into and out of the Levant.

CHAPTER 16

The Religions That Shaped the Modern Levant

It is all about those olive trees, plain and simple

THE EVENTS OF THE LAST two millennia have left the Levant fractured into somewhat autonomous and sometimes isolated entities. Religion played a major part in keeping each entity together, and in a place like Lebanon, these entities turned into communities, where religion was the primary element of identity.

Lebanon is nearly the same size as the island of Cyprus, one fiftieth the size of Spain and one sixty-fifth the size of Texas. Despite its small geographical footprint, it is home to eighteen different religious affiliations or sects. The Lebanese population, uniquely composed of many distinct religious communities, is perfectly suited for an investigation of the impact of religion on the evolution of genetic patterns through time. Genetic differentiation between the various communities in Lebanon is clearly observed and dates to at least one thousand years before the establishment of Christianity and Islam. All DNA analyses to date indicate that the differentiating patterns observed in the various communities in Lebanon started with the Achaemenid Persian conquest of the Levant (c. 530 BCE). Genetic differentiation between the Christian and Muslim communities in today's Lebanon predates the foundation of Christianity, as per the Y chromosome data.

These findings are not a bit surprising, considering various Lebanese communities have been derived from a common genetic stock since the Neolithic Period, as we have seen. From the Last Glacial Period expansion onward, these communities most likely differentiated due to relative geographical isolation and drift. On the one hand, some of these communities were genetically impacted as they mixed with the groups that occupied the Levant since the middle of the first millennium BCE.

On the other hand, some communities were already well established, isolated, and completely independent, like the Maronites of Mount Lebanon, where no evidence of substantial recent admixture is observed. This community was somewhat genetically differentiated from the rest due to long periods of isolation prior to adopting Christianity, which arrived from the north with the Maronite monks fleeing Syria in the late sixth century CE.

In the mid-seventh century CE, other communities, primarily along the Lebanese coast, adopted Islam after the Arab conquest of the Levant that began in 634 CE under Caliph Omar Ben Khattab. In other words, the various (religious) Lebanese communities had evolved in situ from the same genetic stock (Natufians, Neolithic Anatolians, and Neolithic Iranians), and established themselves in various geographical regions, adopting similar Levantine cultures and traditions. Then religions came upon them and stratified them and differentiated them further, both genetically and culturally.

TWO THOUSAND YEARS OF RELIGIOUS CLUSTERING

Analyzing the genomes of individuals belonging to different religious communities in Lebanon reveals a stark level of differentiation. The main religious groups—Christians, Muslims, and Druze—can be easily distinguished based on DNA analysis. These three groups must have accumulated enough genetic differences from one another that they can be easily distinguished when plotted on a graph representing the results of principal component analysis (PCA).

PCA is a nice visual tool (geneticists use it often) to identify groups of individuals or populations who are genetically distant from each other after accumulating enough genetic differences. Individuals or populations that have accumulated enough genetic differences (mutations) tend to show more separation, yielding distinct clusters. The three clusters shown in the following graph are the three main religious groups in Lebanon: Muslims, Christians, and Druze. When populations cluster away from each other, as depicted in the graph, this does not mean that they have no genetic similarities; it simply means that there are enough detectable differences to be able to separate them. As a matter of fact, all three clusters share substantial common genetics, and all the DNA analyses demonstrate that these three groups started to differentiate from a common source population around 2,000 years ago.

Populations who later adopted Christianity were the first (oldest) group to differentiate from the other two groups,

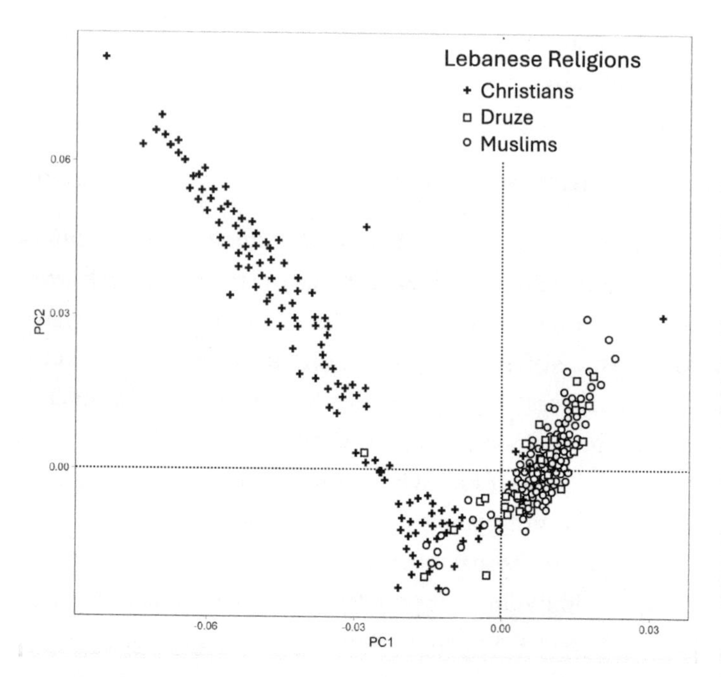

Principal component analysis shows clustering by religion of the Lebanese population. Each symbol represents one individual.

around 400 BCE. This period corresponds to the decline of Phoenicia, control of the region by Hellenistic rulers, and the foundation of Christianity, indicating a relative isolation of this group for more than two thousand years, the longest among the three groups.

Some level of mixing between Lebanese Muslim communities and populations with sub-Saharan ancestry continued up until roughly 650 years ago. That time period corresponds to the Ottoman Empire's dominance and the appearance of a semi-autonomous state in Lebanon. These dates suggest that the Levantine populations probably maintained contact with other populations in the region carrying the sub-Saharan genes until they adopted different cultures or got separated by rising political states.

Two major conclusions can be drawn from the graph. First, the religious clustering can be explained only by the endogamous practices observed in these three religious groups. People tend to marry within their religion, partly for their practices and faith but also because, as in Lebanon, communities are often geographically segregated based on religion. When the same genetic analyses were conducted on the same Lebanese communities but based on geographical affiliation and not religion, no distinct clusters could be identified, leading one to conclude that the genetic clustering is based on religious affiliation and not geography.

Second, conversion between religions has not been a significant practice (almost no crosses among the circles and very few circles among the crosses on the graph). It is surprising, given the multiple waves of occupation, each dominated by a different religion, that have ruled this small country. Having said that, one cannot discount reconversions, which have been documented in several communities throughout the Levant. The genetic clustering that is observed here indicates how culture and in particular religion can impact the gene pool.

THE DRUZE ARE NOT A DISTINCT POPULATION

The Druze were isolated from the other two groups around the time of the development of the Druze faith

(c. 1017 CE) and their split from other Muslims, mostly the Shia Muslims.

The Human Genome Diversity Project (HGDP) classified the Druze as a population. The HGDP is a study that was conducted by a group of scientists from Stanford University in collaboration with the French Centre d'Étude du Polymorphisme Humain, starting in the 1990s. Scientists collected DNA from individuals belonging to more than fifty different populations in Africa, Europe, Asia, Oceania, and the Americas. The target populations or groups were initially defined as those who were isolated and likely to disappear as distinct groups in the future due to population expansions. The study's target in a nutshell was "genetic isolates," if such populations still existed. The purpose of the study was to "understand genetic diversity in human populations." Studying the genetic signatures of these isolated populations was paramount to documenting this diversity before it got lost through admixture.

From the start, the project received a lot of criticism from the indigenous populations, who saw it as more interested in documenting diversity than in trying to protect it. The scientists leading the project had to redefine their population recruitment strategy, seeking the collaboration of anthropologists and actively engaging the indigenous communities targeted for recruitment. While this project was well intentioned and resulted in many seminal publications that revolutionized the field of population genetics, it left a sour aftertaste among indigenous communities throughout the world. When several years later we launched the Genographic Project, we learned from the mistakes made by the

HGDP. Even though we engaged indigenous communities early on, we still faced considerable resistance from many indigenous groups.

Druzism is a faith, and people who follow this faith are called Druze. This faith was founded by Hamzah Ibin Ali during the reign of the Fatimid caliph Al-Hakem Bi Amr Allah in Egypt around 1000 CE. Hamzah Ibin Ali appointed five propagators for this new faith. One of the most prominent was Muhammad Bin Ismail Al Darazi, who spread the faith in the mountains of the Levant. The people of this faith came to be called "Druze" after him, though they prefer to be called "unitarians" (Muwahhideen), in reference to their monotheist faith. After the disappearance of Caliph Al-Hakem under mysterious circumstances, his successor, Al-Zaher, and Al-Hakem's sister ruthlessly persecuted the Druze and forced them out of Egypt. To escape persecution, the followers of this faith took refuge in the mountains of the Levant, where many sympathizers from the local Levantine communities joined their faith. Proselytization to Druzism stopped in the fourteenth century CE and conversion has not been allowed since. Today an estimated one million people of the Druze faith call the Levant their home. They are concentrated in the Chouf area of Mount Lebanon, Houran Mountain in Syria, and mountainous areas west and north of Amman, Jordan, and the Galilee.

DNA work has shown that the Druze, like the other Lebanese religious communities, are derived from the Lebanese population at large. They show some expected genetic structure due to endogamy and geographical isolation, but they share very little of that structure with other Druze from the region. This faith was adopted by the local inhabitants of the Lebanese mountains around the eleventh century. The Druze are a group of a common faith and are not an "ethnic population," and it is absurd that they are referred to as a population in the current published literature.

RELIGIOUS CLUSTERING AND HAPLOGROUP DISTRIBUTION

Specific paternal genetic lineages (Y haplogroups) became more associated with certain religious groups than others, and some of the Y haplogroup distributions revealed important facts about the history of some of these religious groups.

THE GENETIC IMPACT OF THE CRUSADERS

After the various communities in Lebanon differentiated, they were further exposed to varying levels of genetic admixtures that were also driven by religion. The following is one such example.

Since the Western European origin of the Crusaders is well documented, one can theoretically look for Western European genetic signatures among the current inhabitants of the Levant to figure out whether they descended from the Crusaders, or so it was thought. This idea seemed in fact very simple for another genetic study: investigate the Y chromosome haplogroups and haplotypes among the modern Levantine population, and see whether some haplogroups are European specific.

The study revealed that the European-abundant R1b haplogroup was present in about 6 percent of the Lebanese population. Further, the exclusive Northwestern European haplogroup I was also found in the population, albeit at a lower frequency. The relatively high frequency of R1b does not necessarily mean that all the R1b haplogroups that were found in the Levant are from Crusaders, because R1b is also present in Turkey and other parts of Central and Eastern Europe, and the genetic impact of these populations in the Levant is well documented. The study's assumption at this point was that R1b could have come from Europe with the Crusades. But how could it be ruled out that it did not come during the Greek, Roman, Persian, Byzantine, or Ottoman occupation of the Levant, any of which could have been an R1b source population?

The observation of the I haplogroups (almost exclusive to Northwestern Europe) in the Levantine population provided further evidence that what was being detected here was some of the Crusaders' haplogroups after all. To test this assumption, the haplotypes of all the R1bs in the Lebanese population were further investigated to determine their origins. Among the different haplotypes that were found in the Lebanese R1bs, some had Eastern European origins, while others were from West Asia. But the most common haplotype was restricted to Western Europe. It was found in twenty-six European populations at high frequencies, but not even one of them was found east of Hungary. One may argue that this haplotype could have originated independently in the Levant and not through migration. If this was the case, one would expect to find it across the Levant, albeit at a low frequency. More than one thousand individuals from the Levant were tested, and not even one carried this European haplotype. Hence the conclusion was that the only possibility for this haplotype to reach this high a frequency and to be geographically restricted was by way of migration from Western Europe. It was given the name WES haplotype for Western European Specific.

The WES haplotype was most common among Lebanese Christians, although other religious communities carried it as well at lower frequencies. This was the first example of using genetic testing to show that religion-driven movement had an impact on the genetic makeup of the Levant. The high frequency of the WES haplotype can be easily explained; many of the Crusaders remained in the Levant and did not go back to Europe. Some of them eventually converted to other religions.

A HAPLOGROUP DISTRIBUTION THAT FOLLOWS A RELIGIOUS LINE

There is a pattern for a haplogroup distribution that follows a specific religious line, that of the Maronites of Mount Lebanon. The Y chromosome L1b haplogroup shows differential frequencies among the various religious groups of Lebanon. This signature is exclusively present in the Maronites of Mount Lebanon and not present in any of the other religious groups tested. The distribution of this haplogroup was surprising as it is not present anywhere else in the Levant.

My interest in this haplogroup goes beyond my scientific curiosity, I happen to belong to this haplogroup, and so do

10 percent of the people in my hometown. In the Levant, the distribution of this haplogroup is almost exclusively Maronite, and it is present only in Mount Lebanon. This distribution restriction indicates that the signature has remained isolated and could not have been present for long in the population; otherwise we would have observed at least some expansion into neighboring regions or other religious communities, unless its isolation had been extremely and unusually complete, which was the case.

WHO ARE THE MARONITES?

The Maronites, who live primarily in Mount Lebanon, are a group that adopted Christianity in the late fifth century CE. Their name is derived from a monk named Maroun, who lived in seclusion as a hermit on the left bank of the Orontes River in Emesa, modern-day Homs in Syria. After his death in 410 CE his followers, to commemorate his life, erected a monastery in the place where he had lived. They were devout in their Catholic Orthodox faith, fully accepting and adopting the Council of Chalcedon decree of 451: "We teach . . . one and the same Christ, Son, Lord, Only-begotten, known in two natures, without confusion, without change, without division, without separation."

When the Monophysites, who believed in the solely divine nature of Christ, gained power in Byzantium, they persecuted the Maronite monks. The Maronites fled to Mount Lebanon, where they preached their faith. They soon became a dominant force in Mount Lebanon and appointed their own patriarch, distinguishing themselves and somewhat defying Byzantine rule. Byzantium did not approve of this separation and in 694 CE launched an attack on the Maronites of Mount Lebanon to subdue them. The two fought ferociously in Amioun, in northern Lebanon. Since that time, the Maronites have occupied Mount Lebanon with full autonomy, relying for their survival on agriculture, which they have skillfully mastered. During the Islamic expansions and the Byzantine Arab wars, the Maronites managed to hold on to their territories with minimal harm.

Around 670 CE—during the reigns of Constans II, the Byzantine emperor, and Mu'awiyah, the first caliph of the Umayyad Caliphate—a group called the Mardaites arrived in the Lebanese mountains, where they integrated with the local inhabitants, the Maronites. The reason for their arrival is not clear; some sources state that they were recruited from eastern Turkey or Armenia by the Byzantine army in its fights against Mu'awiyah and the Arabs. Other sources claim that they rebelled against Constans II in support of Saborius, a Byzantine general of Armenian descent. After the accidental death of Saborius, they remained an entity on their own with some autonomy under Byzantine rule. What most historians agree on, however, is that the Mardaites arrived in the mountains of Lebanon around 670 CE and were instrumental in helping Justinian II secure a Byzantine grip on the northern Levant against the Arabs. They constituted a force to be feared by the Byzantines and the Arabs, and both wanted to pacify them. Mu'awiyah had to pay a tax to the Mardaites to avoid their hostilities.

It is therefore most plausible that the Mardaites may have been the source of the L1b haplogroup observed in the

Maronites of Mount Lebanon. In particular, the subclade (subbranch) of this haplogroup in the Lebanese Maronites is L1b-M317, and except for the Maronites of Mount Lebanon, it is present at an extremely low frequency across the Levant but is very prevalent in eastern Turkey and Armenia, reaching more than 20 percent in certain parts. Out of the several thousand samples analyzed from Southwest Asia, only the Maronites of Mount Lebanon had this subclade present. This information supports the hypothesis that the Mardaites, had they been the source population for this haplogroup, would still constitute a substantial component of today's Maronite genetic makeup.

However, the L1b story turns out to be a bit more complicated than can be explained by the Mardaite theory alone. Additional genetic analyses revealed that while the L1b-M317 branch that is specific to the Maronites of Mount Lebanon did indeed split from a sister branch that is located in eastern Turkey, this split dates back several thousand years, well beyond 667 CE, the date of the arrival of the Mardaites to Mount Lebanon. This means that the L1b Maronites must have been occupying the mountains of Lebanon well before the time of the Mardaites and for that matter well before the arrival of Christianity to Mount Lebanon.

These new findings are compelling, as they tell the story of the people who may have been the first inhabitants of the mountains of Lebanon. They provide not only a possible date of arrival but also the potential source population (Armenian). The archaeological evidence recovered so far from various spots in Mount Lebanon provides dates that are consistent with the L1b findings, all suggesting a start of community life in the Lebanese mountains around the Early Bronze Age (5,000 to 4,100 years ago).

What is surprising about these DNA findings is that one could still distinguish an isolated community through its paternal lineage in a region where so much human mobility and human interaction have occurred. What is even more surprising is that this genetic isolation is also apparent in the autosomal DNA of this community. Y lineages can still be detected through many generations since the Y chromosome does not recombine through admixture, but autosomal DNA gets shuffled every generation. The Maronites

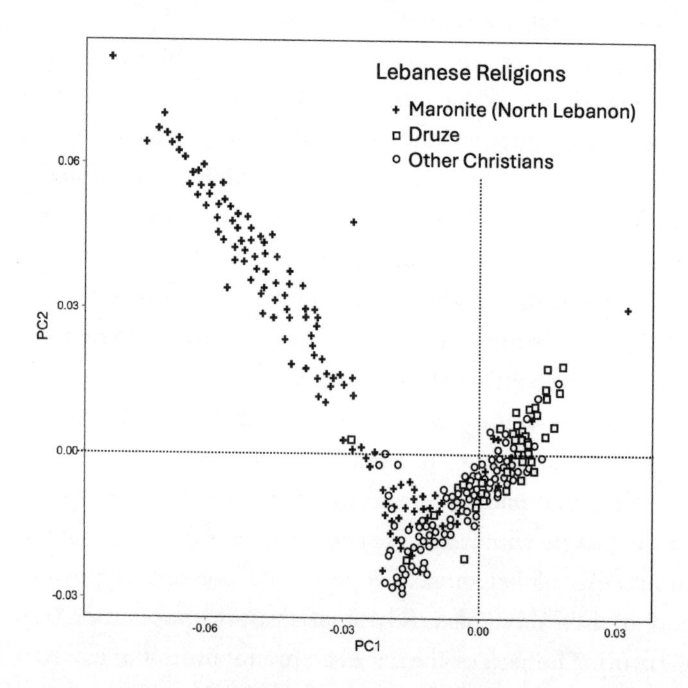

Clustering of the Maronite community in northern Mount Lebanon

of Mount Lebanon show a distinctive genetic cluster from all other communities in Lebanon, including other Maronites who are not from Mount Lebanon.

WHAT BROUGHT PEOPLE FROM THE CAUCASUS TO MOUNT LEBANON?

Approximately 8,200 years ago, a significant climatic event led to widespread drought in the Levant, most of Europe, and Southwest Asia. With the return of better climate conditions several centuries later, people from the Caucasus region expanded across Asia. The Maronite L1b group is thought to have migrated from the Caucasus approximately 5,000 years ago, along a path that led it to the Levant. This migration from the Caucasus was likely prompted by the improvement in climate, as individuals sought more favorable habitats in the northern Levant. The northern Lebanese mountains, characterized by their hilly topography, proved to be an ideal habitat for these travelers from the Caucasus, offering numerous shelters and lush, verdant areas with abundant precipitation.

These mountains have continued to this day to provide favorable, indeed ideal, conditions and necessary resources for uninterrupted habitation. People have been residing in the mountains of northern Lebanon since at least the Early Bronze Age, as supported by archaeological discoveries and now by DNA evidence. These inhabitants were most likely the first inhabitants of the Lebanese mountains and the first to adopt the Maronite faith that was brought by the disciples of Maroun from Syria around the late fifth century CE. They remain somewhat isolated even today, and they have always been described as resilient against occupiers and invaders.

With the exception of two individuals from Sicily and one individual from Libya, all the L1bs in the study were found in the mountains of Lebanon and belonged to the same Maronite community. The Libyan L1b is most likely a descendant of a Phoenician sailor, as Libya has constituted a major settlement for the Phoenicians since the early first millennium BCE.

While the two Sicilian individuals could be the direct descendants of Phoenician sailors, a more plausible explanation exists. In 1584 the Maronite College in Rome was established to create a cultural link between the Christians of the Levant and Rome. This college was founded by a Maronite patriarch from the northern Lebanese mountains. Active for 228 years, the college played a key role in exchanging and promoting cultural practices between the two sides of the Mediterranean, and most of these practices are still observed today.

The college served as an educational conduit for several hundred male disciples of all age groups and from various geographical locations. Many of them came from neighboring Syria and Cyprus. The majority of these disciples would frequently travel to Italy from northern Lebanon, often passing through Sicily. Indeed, Sicily served as a main site for connecting the Maronite community to Europe in the sixteenth and seventeenth centuries. A significant number of disciples from the northern Lebanese mountains were among those associated with the Maronite College, and quite plausibly the two Sicilian Levantine L1bs arrived in Sicily as part of a cultural exchange initiative between Rome and the Christians of the Levant. One of the two Sicilians happens to carry the same expanded Y haplotype as me, with one single change from a perfect match—most likely a distant cousin!

THE IMPACT OF THE ISLAMIC EXPANSION

Haplogroups J1 and J2 vary significantly in some of the major religious communities in Lebanon. The J1 frequency among Lebanese Muslims is 25 percent while only 15 percent in non-Muslims, while the J2 frequency is more than 30 percent among Christians and less than 20 percent in Muslims. Haplogroup J1 and more precisely J1-P58 is the most common haplogroup in Arabia and most likely spread with the Islamic expansion. It is not surprising to see that haplogroup J1 and precisely J1-P58 is more dominant in the Levantine Muslim population. The different J1 frequencies between Muslims and non-Muslims could have resulted from a genetic drift and not from the Islamic expansion. The genetic data to date, however, strongly indicate that such differences could have come only from external genetic input that occurred forty-two generations ago, the time of the Islamic expansion.

THE ETHNICITIES AND IDENTITIES IN THE LEVANT

Do not look for DNA to tell you who you really are and where you belong. It is a fallacy!

It is populations, not individuals, within certain geographical boundaries, that are assigned certain genetic signatures. They become commonly referred to as "ethnicities." These populations may have previously been a collection or a mix of many smaller, genetically distinct subpopulations that amalgamated (through mating) and adopted a geographical location as a home. Populations are not static. Hence ethnicities are not static; they mix, they change, they move, and they evolve. As they evolve, these populations adopt certain habits and become culturally distinct. These habits and cultures, like genetic signatures, become associated with ethnic backgrounds.

Identities are outward projections rooted in memorable imageries, while cultures are habits that are collectively adopted by a group of individuals who decided together to occupy a geographical space in a time period. DNA is the mesh upon which the history of how humans spread from a single population of hunters and gatherers out of Africa, then branched into a variety of distinctive subpopulations with diverse cultures, is woven. DNA does not define or determine an identity, a culture, or an ethnicity. It uncovers stories from the past but does not reveal who we really are. Identity is a rich and complex concept. It is the origin, the birthplace, and the early interactions. It is the lifetime experience. Identity is a personal and unique identifier; identity is yours to make. Cherish it, be proud of it, and don't be shy about sharing it with the world.

DON'T CALL THEM MINORITIES

There are no religious or ethnic minorities in the Levant. There are only groups and communities that have occupied the Levant uninterrupted for thousands of years. They established small dwellings first, using mud, wood, and stones. They hunted and farmed the land. They established the first communities and worshipped their gods. Around 8,000 years ago they started building towns. Byblos, eleven kilometers from where I am writing these lines, has been continuously inhabited ever since. This is where the first complete alphabet as we know it was carved on a sarcophagus. The people spoke Canaanite, Aramaic, Syriac, Lebanese, and Arabic. They developed habits, rituals, and cultures; built structures; and made good wine and olive oil. They were attacked and forced to sail across the Mediterranean in search of metals and other precious materials for their aggressors.

They never fled. They never left their land.

The Levantines possess an eight-thousand-year rich heritage, and most share a similar genetic ancestry. No unique culture defines them or ever did. Cultures move like people, get transmitted like genes, and some unfortunately disappear, like languages. The Levant is a land of many cultures where some evolved, some morphed into other cultures, some migrated, some remained isolated, and some entirely disappeared. All the many remaining cultures of the Levant tell a Levantine story, and it is the collection of these stories that define the Levant today, the land of a thousand cultures.

ACKNOWLEDGMENTS

I AM GRATEFUL TO MY FRIENDS Lisa Matisoo-Smith, Magda Boudagher Kharrat, Samir Melki, and François Kamar, who encouraged me from the outset and provided me with judicious guidance throughout the writing process. They read and commented on the early drafts and read every chapter (well, almost every chapter) and helped me significantly improve the manuscript. Thank you, Magda, for the long and many hours you dedicated to creating the illustrations for this manuscript.

I am grateful to David Sheehan and Niall and Deirdre Jones for their multiple readings of the later drafts and making valuable comments and edits.

I am indebted to my dear friend Rami (Hiram) Corm for his fatherly guidance and for rubbing so much of his wisdom into the manuscript. Some of it stuck, I am certain.

I had such a wonderful experience working with my editor, Marie Pantojan, who helped me reshape the manuscript presentation and gave it a more appealing profile. I am thankful to Azraf Khan for guidance and valuable insights that were impactful. I am very grateful to Ben Greenberg for believing in this project.

I am profoundly grateful to my late father-in-law for always showing me the way.

To Nassim Nicolas Taleb, my role model and mentor, I am eternally grateful. Without you, this manuscript would never have seen the light. Thank you, Nassim, for your mentorship. I am so lucky to have you not only as a mentor but as a friend.

I want to express my gratitude to my sister and brother, who are a perpetual source of stability and pride for me. I thank them for their unconditional support, trust, and encouragement, which have guided me not just during the writing process but throughout my life. I am so grateful for all the sacrifices they made so I could have a career that I love. I owe it to you both.

To my late parents, I am at a loss for words; not a single night passes by without me thinking how lucky I am to have had you. You showed me the true meaning of pure and altruistic love.

I am fortunate to have an amazing partner who has consistently showered me with love, care, and encouragement since the day we met twenty-five years ago. Aline, my most unwavering supporter, believed in me during both the highs and lows and motivated me with resolute confidence, without hesitation or doubt, to see this project to a successful completion.

I am blessed with two wonderful daughters, Yasha and Pia, two stars who illuminate my life and inspire me to become a better person every day. They are my constant source of pride and joy. They are my most serious critics. Both added their magical touches to the introduction to the book, improved it tremendously, and made me so proud in the process.

Finally, I dedicate this book to Louli, the brother who was violently and prematurely torn off from our family. He was our guiding beacon and our anchor to safety, and we miss him dearly. I wanted him so much to see this book in print.

NOTES

CHAPTER 1: ORIGINS AND IDENTITIES

5 **German historian August Ludwig von Schlözer** J. G. Eichhorn, "Bemerkungen über den Text des Propheten Jeremias," *Repertorium für biblische und morgenländische Litteratur 1777*; Han F. Vermeulen, "Early History of Ethnography and Ethnology in the German Enlightenment: Anthropological Discourse in Europe and Asia, 1710–1808," PhD diss., University of Leiden, 2008.

5 **designate both peoples and languages** Moshe Zimmermann, *Wilhelm Marr: The Patriarch of Anti-Semitism* (New York: Oxford University Press, 1986).

6 **"*Roots* is not a word"** Amin Maalouf, *Origins: A Memoir* (New York: Farrar, Straus & Giroux, 2009).

8 **"Just like the ancient Greeks"** Amin Maalouf, *In the Name of Identity: Violence and the Need to Belong* (1998; reprint New York: Time-Warner, 2001).

10 **Hay, or Haya, is the name** Simon Payaslian, *The History of Armenia: From the Origins to the Present* (Basingstoke: Palgrave Macmillan, 2008).

13 **argues fervently against** Samuel P. Huntington, *The Clash of*

Civilizations and the Remaking of World Order (New York: Simon & Schuster, 1996).

CHAPTER 2: ANCESTRY OR HERITAGE?

16 **the concept of race** Carl Linnaeus, *Systema Naturae* (1735; reprint London: British Museum, 1956).

17 **based on cranium measurements** Johann Friedrich Blumenbach, *On the Natural Variety of Mankind,* trans. T. Bendyshe, in Blumenbach, *The Anthropological Treatises of Johann Friedrich Blumenbach* (London: Longmans Green, 1865), 145–276.

17 **phenotypic (morphological) features** François Bernier et al., *Linnaeus and de Maupertuis on Race* (Bristol: Thoemmes, 2003).

18 **"race" reached its pinnacle** Claude Lévi-Strauss, *Race and History* (Paris: UNESCO, 1952).

18 **race means a great deal** Nadia Abu El-Haj, "The Genetic Reinscription of Race," *Annual Review of Anthropology* 36 (2007).

20 **The term *Indian*** Michael A. Peters and Carl T. Mika, "Aborigine, Indian, Indigenous or First Nations?" *Educational Philosophy and Theory* 49, no. 13 (2017): 1229–34.

21 **"Since culture is nothing but"** Fredrik Barth, *Ethnic Groups and Boundaries* (1969; reprint London: G. Allen & Unwin, 1970).

21 **"Someone is a Lue"** Michael Moerman, "Ethnic Identification in a Complex Civilization: Who Are the Lue?" *American Anthropologist* 67 (1965): 1215–30.

22 **"Ethnic distinctions do not depend"** Barth, *Ethnic Groups and Boundaries*.

23 **population replacement or genetic discontinuity** Svante Pääbo, "The Mosaic That Is Our Genome," *Nature* 421 (2003): 409–12; David Reich, *Who We Are and How We Got Here: Ancient DNA and the New Science of the Human Past* (New York: Pantheon, 2018).

24 **The exact origin of the Beaker people** Iñigo Olalde et al.,

"The Beaker Phenomenon and the Genomic Transformation of Northwest Europe," *Nature* 155 (2018): 190.

CHAPTER 3: FROM AFRICA TO THE LEVANT

29 **oldest *Homo sapiens* skull** Javier R. Luis et al., "The Levant Versus the Horn of Africa: Evidence for Bidirectional Corridors of Human Migrations," *American Journal of Human Genetics* 74, no. 3 (2004): 532–34.

29 **first archaeologically documented migration** Saioa López, Lucy van Dorp, and Garrett Hellenthal, "Human Dispersal out of Africa: A Lasting Debate," *Evolutionary Bioinformatics Online* 1, no. 2 (suppl. 2) (2015): 57–68; Christopher Brian Stringer and Peter Andrews, "Genetic and Fossil Evidence for the Origin of Modern Humans," *Science* 239, no. 4845 (1988): 1263–68; Li Jin and Bing Su, "Natives or Immigrants: Modern Human Origin in East Asia," *Nature Reviews Genetics* 1, no. 2 (2000): 126–33; Simon J. Armitage et al., "The Southern Route 'Out of Africa': Evidence for an Early Expansion of Modern Humans into Arabia," *Science* 331, no. 6016 (2011): 453–56.

29 **cave in Jebel Irhoud** Jean-Jacques Hublin et al., "New Fossils from Jebel Irhoud, Morocco and the Pan-African Origin of Homo Sapiens," *Nature* 558 (2017): 289–92.

30 **Apidima Cave in southern Greece** Katerina Harvati et al., "Apidima Cave Fossils Provide Earliest Evidence of Homo Sapiens in Eurasia," *Nature* 571 (2019): 500–4.

30 **successful and permanent establishment** Eleanor M. L. Scerri et al., "Did Our Species Evolve in Subdivided Populations Across Africa, and Why Does It Matter?" *Trends in Ecology and Evolution* 33, no. 8 (August 2018): 582–94; Nick Drake and Paul Breeze, "Climate Change and Modern Human Occupation of the Sahara from MIS 6-2," in Brian Stewart and Sacha Jones, eds., *Africa from MIS 6-2: Population Dynamics and Paleoenviron-*

ments (New York: Springer, 2016); Chris Stringer, *Lone Survivors: How We Came to Be the Only Humans on Earth* (New York: St. Martin's Griffin, 2013).

30 **left the African continent** Elena A. A. Garcea, ed., *South-Eastern Mediterranean Peoples Between 130,000 and 10,000 Years Ago* (Oakville, CT: David Brown Book Co., 2010); Naama Goren-Inbar and John D. Speth, *Human Paleoecology in the Levantine Corridor* (Oxford: Oxbow Books, 2004); Yuval Noah Harari, *Sapiens: A Brief History of Humankind* (New York: HarperPerennial, 2018).

31 **the climate in Africa shifted** Chris Stringer and Robin McKie, *African Exodus: The Origins of Modern Humanity* (New York: Henry Holt, 1998).

31 **abundant archaeological evidence** Huw S. Groucutt et al., "Homo sapiens in Arabia by 85,000 Years Ago," *Nature Ecology and Evolution* 2, no. 5 (2018): 800–9; Erella Hovers, *The Lithic Assemblages of Qafzeh Cave* (New York: Oxford University Press, 2009); Ofer Bar-Yosef, *Pastoralism in the Levant: Archaeological Materials in Anthropological Perspectives* (Madison, WI: Prehistory Press, 1992).

31 **Skhul and Qafzeh caves** Christopher Brian Stringer, R. Grün, Henry Schwarcz, and Paul Goldberg, "ESR Dates for the Hominid Burial Site of Es Skhul in Israel," *Nature* 338 (1989): 756.

32 **Tabun, Kebara, and Geula caves** Richard G. Klein, "Anatomy, Behavior, and Modern Human Origins," *Journal of World Prehistory* 9, no. 2 (1995): 167–98.

33 **to return to Africa** Goren-Inbar and Speth, *Human Paleoecology in the Levantine Corridor.*

33 **very few sites in Southwest Asia** John R. Stewart and Christopher Brian Stringer, "Human Evolution out of Africa: The Role of Refugia and Climate Change," *Science* 335, no. 6074 (2012): 1317–21.

34 **Reconstructed climate maps** Daniel Richter et al., "The Middle

to Upper Palaeolithic Transition in the Levant and New Thermoluminescence Dates for a Late Mousterian Assemblage from Jerf-al Ajla Cave (Syria)," *Paléorient* 27, no. 2 (2001): 29–46; Anne-Lise Develle et al., "A 250ka Sedimentary Record from a Small Karstic Lake in the Northern Levant (Yammoûneh, Lebanon): Paleoclimatic Implications," *Palaeogeography, Palaeoclimatology, Palaeoecology* 305, no. 1 (2011): 10–27; H. Cheng et al., "The Climate Variability in Northern Levant over the Past 20,000 Years," *Geophysical Research Letters* 42, no. 20 (2015): 8641–50.

34 **mostly arid Levant** Drake and Breeze, "Climate Change and Modern Human Occupation"; Miryam Bar-Matthews, Avner Ayalon, and Aaron Kaufman, "Late Quaternary Paleoclimate in the Eastern Mediterranean Region from Stable Isotope Analysis of Speleotherms at Soreq Cave, Israel," *Quaternary Research* 47, no. 2 (1997): 155–68; Arie S. Issar and Mattanyah Zohar, *Climate Change: Environment and History of the Near East,* 2nd ed. (New York: Springer, 2007).

34 **major cultural shift** Lara Hajar, Maya Haïdar-Boustani, Carla Khater, and Rachid Cheddadi, "Environmental Changes in Lebanon During the Holocene: Man vs. Climate Impacts," *Journal of Arid Environments* 74, no. 7 (2010): 746–55; Colin Renfrew and Paul Bahn, *Archaeology: Theories, Methods, and Practice* (London: Thames & Hudson, 2016).

35 **more refined stone tools** Hajar et al., "Environmental Changes in Lebanon During the Holocene"; John J. Shea, "Behavioral Differences Between Middle and Upper Paleolithic *Homo sapiens* in the East Mediterranean Levant: The Roles of Intraspecific Competition and Dispersal from Africa," *Journal of Anthropological Research* 63, no. 4 (2007): 449–88; Dafna Langgut et al., "Evidence for a Humid Interval at ~56–44 Ka in the Levant and Its Potential Link to Modern Humans Dispersal out of Africa," *Journal of Human Evolution* 124 (November 2018): 75–90.

35 **descendants of a Neanderthal and *Homo sapiens* mix** Mathias

Currat and Laurent Excoffier, "Strong Reproductive Isolation Between Humans and Neanderthals Inferred from Observed Patterns of Introgression," *Proceedings of the National Academy of Sciences* 108, no. 37 (2011): 15129–34; Ewen Callaway, "Modern Human Genomes Reveal Our Inner Neanderthal," *Nature* (January 29, 2014); Alan R. Rogers, Ryan J. Bohlender, and Chad D. Huff, "Early History of Neanderthals and Denisovans," *Proceedings of the National Academy of Sciences* 114, no. 37 (2017): 9859–63.

35 **ideal place for it** Aaron Jonas Stutz, "Near East (Including Anatolia): Geographic Description and General Chronology of the Paleolithic and Neolithic," in C. Smith, ed., *Encyclopedia of Global Archaeology* (New York: Springer, 2014), 5182–208.

35 **The Ksar Akil cave** Marjolein D. Bosch, Amy L. Prendergast, and Jean-Jacques Hublin, "New Chronology for Ksâr 'Akil (Lebanon) Supports Levantine Route of Modern Human Dispersal into Europe," *Proceedings of the National Academy of Sciences* 112, no. 25 (2015): 7683–88.

35 **warming and moistening** Richter et al., "Middle to Upper Palaeolithic Transition in the Levant"; Jed O. Kaplan et al., "Large Scale Anthropogenic Reduction of Forest Cover in Last Glacial Maximum Europe," *PLoS ONE* 11, no. 11 (2016): e0166726.

35 **modern humans expanded** Bar-Matthews, Ayalon, and Kaufman, "Late Quaternary Paleoclimate in the Eastern Mediterranean Region"; Ofer Bar-Yosef, "Climatic Fluctuations and Early Farming in West and East Asia," *Current Anthropology* 52, no. 2 (2011): S175–93.

35 **collection of seeds and wild cereals** Joanne Clarke, ed., *Archaeological Perspectives on the Transmission and Transformation of Culture in the Eastern Mediterranean* (Oxford: Oxbow, 2005).

36 **small community way of life** Bar-Yosef, "Climatic Fluctuations and Early Farming"; Ofer Bar-Yosef and A. Belfer-Cohen, "From Foraging to Farming in the Mediterranean Levant," in

A. B. Gebauer and T. D. Price, eds., *Transitions to Agriculture in Prehistory* (Madison, WI: Prehistory Press, 1992), 21–48.

CHAPTER 4: THE DNA TRAIL

42 **two migratory events** Eleanor M. L. Scerri et al., "Did Our Species Evolve in Subdivided Populations Across Africa, and Why Does It Matter?" *Trends in Ecology and Evolution* 33, no. 8 (August 2018): 582–94; Chris Stringer and Robin McKie, *African Exodus: The Origins of Modern Humanity* (New York: Henry Holt, 1998); Max Ingman, Henrik Kaessmann, Svante Pääbo, and Ulf Gyllensten, "Mitochondrial Genome Variation and the Origin of Modern Humans," *Nature* 408 (2000): 708–13; I. Tattersall, "Human Origins: Out of Africa," *Proceedings of the National Academy of Sciences* 106, no. 38 (2009): 16018–21; Rasmus Nielsen et al., "Tracing the Peopling of the World Through Genomics," *Nature* 541 (2017): 302–10.

44 **"Scientific Eve"** Rebecca L. Cann, Mark Stoneking, and Allan C. Wilson, "Mitochondrial DNA and Human Evolution," *Nature* 325 (1987): 31–36.

45 **"Out of Africa" mutation** O. Semino et al., "Origin, Diffusion, and Differentiation of Y-Chromosome Haplogroups E and J: Inferences on the Neolithization of Europe and Later Migratory Events in the Mediterranean Area," *American Journal of Human Genetics* 74, no. 5 (2004): 1023–34.

46 **second migratory phase** J. R. Luis et al., "The Levant Versus the Horn of Africa: Evidence for Bidirectional Corridors of Human Migrations," *American Journal of Human Genetics* 74, no. 3 (2004): 532–44; Luigi Luca Cavalli-Sforza and Marcus W. Feldman, "The Application of Molecular Genetic Approaches to the Study of Human Evolution," *Nature Genetics* 33 (suppl.) (2003): 266–75; Lluís Quintana-Murci et al., "Genetic Evidence

of an Early Exit of *Homo sapiens sapiens* from Africa Through Eastern Africa," *Nature Genetics* 23, no. 4 (1999): 437–41; Q. Fu et al., "A Revised Timescale for Human Evolution Based on Ancient Mitochondrial Genomes," *Current Biology* 23, no. 7 (2013): 553–59; Marjolein D. Bosch, Amy L. Prendergast, and Jean-Jacques Hublin, "New Chronology for Ksâr 'Akil (Lebanon) Supports Levantine Route of Modern Human Dispersal into Europe," *Proceedings of the National Academy of Sciences* 112, no. 25 (2015): 7683–88; L. Pagani et al., "Genomic Analyses Inform on Migration Events During the Peopling of Eurasia," *Nature* 538 (2016): 238–42.

47 **inhabiting the entire African continent** Scerri et al., "Did Our Species Evolve in Subdivided Populations?"

CHAPTER 5: *HOMO SAPIENS* MEET NEANDERTHALS IN THE LEVANT

49 **the Sima de los Huesos cave** L. Slimak et al., "Modern Human Incursion into Neanderthal Territories 54,000 Years Ago at Mandrin, France," *Science Advances* 8, no. 6 (2022): eabj9496.

49 **Several other locations** E. Been et al., "The First Neanderthal Remains from an Open-Air Middle Palaeolithic Site in the Levant," *Scientific Reports* 7, no. 1 (2017): 2958.

49 **diverged sometime in the Middle** S. Horai et al., "Man's Place in Hominoidea Revealed by Mitochondrial DNA Genealogy," *Journal of Molecular Evolution* 35, no. 1 (1992): 32–43.

49 **last common ancestor** Ibid.

50 **initial date of 500,000 years ago** R. Grun et al., "Dating the Skull from Broken Hill, Zambia, and Its Position in Human Evolution," *Nature* 580 (2020): 372–75.

50 **restricts it to the oldest** M. Meyer et al., "Nuclear DNA Sequences from the Middle Pleistocene Sima de los Huesos Hominins," *Nature* 531 (2016): 504.

50 **Neanderthals were shorter** Christopher Brian Stringer and Peter Andrews, "Genetic and Fossil Evidence for the Origin of Modern Humans," *Science* 239, no. 4845 (1988): 1263–68; I. Tattersall, "Human Origins: Out of Africa," *Proceedings of the National Academy of Sciences* 106, no. 38 (2009): 16018–21.

50 **This mixing happened** Svante Pääbo, "The Mosaic That Is Our Genome," *Nature* 421 (2003): 409–12; Svante Pääbo, "The Human Genome and Our View of Ourselves," *Science* 291 (2001): 1219–20; Günter Bräuer and Fred H. Smith, eds., *Continuity or Replacement: Controversies in* Homo sapiens *Evolution* (Rotterdam: A. A. Balkema, 1992); Matthew H. Nitecki and Doris V. Nitecki, eds., *Origins of Anatomically Modern Humans* (1994; reprint New York: Springer, 2013).

51 **at least 1 percent to about 4 percent remnants** R. E. Green et al., "A Draft Sequence of the Neandertal Genome," *Science* 328, no. 5979 (2010): 710–22; B. Vernot and J. M. Akey, "Resurrecting Surviving Neandertal Lineages from Modern Human Genomes," *Science* 343, no. 6174 (2014): 1017–21.

51 **geography- and population-specific** Kay Prüfer et al., "The Complete Genome Sequence of a Neanderthal from the Altai Mountains," *Nature* 505 (2014): 43–49.

51 **Europeans have less** J. D. Wall et al., "Higher Levels of Neanderthal Ancestry in East Asians Than in Europeans," *Genetics* 194, no. 1 (2013): 199–209; K. J. Herrera et al., "To What Extent Did Neanderthals and Modern Humans Interact?" *Biological Reviews* 84, no. 2 (2009): 245–57.

53 **Neanderthal fossils up to 45,000 years ago** Aaron Jonas Stutz, "Near East (Including Anatolia): Geographic Description and General Chronology of the Paleolithic and Neolithic," in C. Smith, ed., *Encyclopedia of Global Archaeology* (New York: Springer, 2014), 5182–208.

53 **survive and thrive** John R. Stewart and Christopher Brian Stringer, "Human Evolution out of Africa: The Role of

Refugia and Climate Change," *Science* 335, no. 6074 (2012): 1317–21.

53 **to confer better immunity** M. Deschamps et al., "Genomic Signatures of Selective Pressures and Introgression from Archaic Hominins at Human Innate Immunity Genes," *American Journal of Human Genetics* 98, no. 1 (2016): 5–21.

56 **at the Neanderthals' expense** Colin Renfrew and Paul Bahn, *The Cambridge World Prehistory: West and Central Asia and Europe* (Cambridge: Cambridge University Press, 2014).

56 **the Denisovans, existed** David Reich et al., "Genetic History of an Archaic Hominin Group from Denisova Cave in Siberia," *Nature* 468 (2010):1053–60.

56 **split from *Homo sapiens*** Ibid.; Stewart and Stringer, "Human Evolution out of Africa."

57 **encountered the Denisovans** Alan R. Rogers, Ryan J. Bohlender, and Chad D. Huff, "Early History of Neanderthals and Denisovans," *Proceedings of the National Academy of Sciences* 114, no. 37 (2017): 9859–63; Prüfer et al., "Complete Genome Sequence"; S. R. Rowning et al., "Analysis of Human Sequence Data Reveals Two Pulses of Archaic Denisovan Admixture," *Cell* 173, no. 1 (2018): 53–61.

57 **all 100 percent human** Luigi Luca Cavalli-Sforza and M. Seielstad, *Genes, Peoples, and Languages* (London: Penguin Books, 2001); Spencer Wells and Mark Read, *The Journey of Man: A Genetic Odyssey* (London: Allen Lane, 2002); Bryan Sykes, *The Seven Daughters of Eve* (New York: Norton, 2001).

CHAPTER 6: OUR EARLY ANCESTORS IN THE LEVANT

64 **Wadi Hilo in southeastern Arabia** M. Uerpmann et al., "HLO1-south: An Early Neolithic Site in Wadi al-Hilo (Sharjah, UAE)," *Arabian Archaeology and Epigraphy* 29, no. 1 (2018): 1–9.

67 **early Natufian culture** Arie S. Issar and Mattanyah Zohar, *Climate Change: Environment and History of the Near East,* 2nd ed. (New York: Springer, 2007); M. A. Zeder, "Domestication and Early Agriculture in the Mediterranean Basin: Origins, Diffusion, and Impact," *Proceedings of the National Academy of Sciences* 105, no. 33 (2008): 11597–604; James Mellaart, *Çatal Hüyük: A Neolithic Town in Anatolia* (London: Thames & Hudson, 1967).

67 **adopted a sedentary lifestyle** James Mellaart, *Earliest Civilizations of the Near East* (London: Thames & Hudson, 1978); M. Özdoğan, "Archaeological Evidence on the Westward Expansion of Farming Communities from Eastern Anatolia to the Aegean and the Balkans," *Current Anthropology* 52 (suppl. 4) (2011): S415–S30; Klaus Schmidt, "Göbekli Tepe, Southeastern Turkey: A Preliminary Report on the 1995–1999 Excavations," *Paléorient* 26, no. 1 (2000): 45–54.

68 **Anatolian and Iranian farmers** Ofer Bar-Yosef and François R. Valla, eds., *The Natufian Culture in the Levant* (Ann Arbor, MI: International Monographs in Prehistory, 1992).

68 **population centers that gave rise** Iosif Lazaridis et al., "Ancient Human Genomes Suggest Three Ancestral Populations for Present-Day Europeans," *Nature* 513 (2014): 409–13.

69 **genetically distinct population** Ibid.

69 **fewer in number** Colin Renfrew and Paul Bahn, *The Cambridge World Prehistory: West and Central Asia and Europe* (Cambridge: Cambridge University Press, 2014); David Rohl, *Legend: The Genesis of Civilisation* (London: Arrow, 1999); Philip K. Hitti, *History of Syria: Including Lebanon and Palestine* (Piscataway, NJ: Gorgias Press, 2004).

70 **Negev highlands, the northeastern Levant** Bar-Yosef and Valla, *Natufian Culture in the Levant*.

70 **referred to as the Harifians** Nigel Goring-Morris, "The Harifian of the Southern Levant," in Ofer Bar-Yosef and François R.

Valla, eds., *The Natufian Culture in the Levant* (Ann Arbor, MI: International Monographs in Prehistory, 1992), 173–234.

CHAPTER 7: FROM ANATOLIA AND THE ZAGROS TO THE LEVANT

73 **The Beldibian culture** James Mellaart, *Çatal Hüyük: A Neolithic Town in Anatolia* (London: Thames & Hudson, 1967); James Mellaart, *Earliest Civilizations of the Near East* (London: Thames & Hudson, 1978); James Mellaart, *The Earliest Settlements in Western Asia: From the Ninth to the End of the Fifth Millennium B.C.* (London: Cambridge University Press, 1967); Arlene M. Rosen and Isabel Rivera-Collazo, "Climate Change, Adaptive Cycles, and the Persistence of Foraging Economies During the Late Pleistocene/Holocene Transition in the Levant," *Proceedings of the National Academy of Sciences* 109, no. 10 (2012): 3640–45.

73 **same time as the Natufians** Ofer Bar-Yosef and François R. Valla, eds., *The Natufian Culture in the Levant* (Ann Arbor, MI: International Monographs in Prehistory, 1992); Charles K. Maisels, *The Emergence of Civilization: From Hunting and Gathering to Agriculture, Cities, and the State in the Near East* (London: Routledge, 2016); Sengül Aydingün, "Early Neolithic Discoveries at Istanbul," *Antiquity* 83, no. 320 (2009).

73 **Neolithic Anatolia contributed** Iosif Lazaridis et al., "Ancient Human Genomes Suggest Three Ancestral Populations for Present-Day Europeans," *Nature* 513 (2014): 409–13; Iosif Lazaridis et al., "Genomic Insights into the Origin of Farming in the Ancient Near East," *Nature* 536 (2016): 419–24.

77 **At sites across the Deh Luran Plain** Ofer Bar-Yosef, "Climatic Fluctuations and Early Farming in West and East Asia," *Current Anthropology* 52, no. 2 (2011): S175–93; Maisels, *Emergence of Civilization;* D. R. Harris, *Origins of Agriculture in Western Central Asia: An Environmental-Archaeological Study* (Philadelphia: University of Pennsylvania Museum of Archaeology and Anthropology, 2010).

77 **Multiple environmental zones** Bernhard Weninger et al., "Climate Forcing Due to the 8200 Cal yr BP Event Observed at Early Neolithic Sites in the Eastern Mediterranean," *Quaternary Research* 66, no. 3 (2006): 401–20.

79 **A remarkable rectangular** Maisels, *Emergence of Civilization.*

79 **corresponds closely to the Fertile Crescent** Ofer Bar-Yosef and A. Belfer-Cohen, "From Foraging to Farming in the Mediterranean Levant," in A. B. Gebauer and T. D. Price, eds., *Transitions to Agriculture in Prehistory* (Madison, WI: Prehistory Press, 1992), 21–48; Mellaart, *The Earliest Settlements in Western Asia;* Weninger et al., "Climate Forcing Due to the 8200 Cal yr BP Event."

79 **referred to these early farmers as Neolithic** J-F. Jarrige, "Mehrgarh Neolithic: New Excavations," in M. Taddei and G. De Marco, eds., *South Asian Archaeology 1997* (Rome: IsIAO, 2000), 259–83.

80 **The basic element** Mellaart, *Earliest Settlements in Western Asia.*

80 **They domesticated cows** M. Nesbitt, "When and Where Did Domesticated Cereals First Occur in Southwest Asia?" in R.T.J. Cappers and S. Bottema, eds., *The Dawn of Farming in the Near East* (Berlin: Ex Oriente, 2002), 113–32.

80 **Samarrans invented irrigation** Mellaart, *Earliest Civilizations of the Near East;* David Rohl, *Legend: The Genesis of Civilisation* (London: Arrow, 1999).

81 **Halaf culture spread** G. Algaze, *The Uruk World System: The Dynamics of Expansion of Early Mesopotamian Civilization* (Chicago: University of Chicago Press, 2005).

CHAPTER 8: THE LEVANT IN THE NEOLITHIC PERIOD

83 **five advantages of the Fertile Crescent** Jared Diamond, *Guns, Germs, and Steel: The Fates of Human Societies* (New York: Norton, 1997), 83.

84 **seasonal movement of these Luristan farmers** Inge Demant Mortensen, *Nomads of Luristan: History, Material Culture, and Pastoralism in Western Iran* (London: Thames & Hudson, 1993).

84 **eight cereal and grain crops** Ofer Bar-Yosef and Avi Gopher, *An Early Neolithic Village in the Jordan Valley* (Cambridge, MA: Peabody Museum of Archaeology and Ethnology, 1997).

CHAPTER 9: POPULATION EXPANSIONS

89 **how they genetically evolved** Daniel E. Platt et al., "Mapping Post-Glacial Expansions: The Peopling of Southwest Asia," *Scientific Reports* 7 (2017): 40338.

89 **underwent long periods of isolation** Ibid.

90 **certain Y chromosome signatures** D. A. Badro et al., "Y-Chromosome and mtDNA Genetics Reveal Significant Contrasts in Affinities of Modern Middle Eastern Populations with European and African Populations," *PLoS ONE* 8, no. 1 (2013): e54616.

92 **Yemenis, for example** Ibid.

98 **reduced level of genetic diversity** M. El-Sibai et al., "Geographical Structure of the Y-Chromosomal Genetic Landscape of the Levant: A Coastal-Inland Contrast," pt. 6, *Annals of Human Genetics* 73 (2009): 568–81.

99 **higher on the Levantine coast** Badro et al., "Y-Chromosome and mtDNA Genetics"; El-Sibai et al., "Geographical Structure of the Y-Chromosomal Genetic Landscape of the Levant."

100 **certain J1 subgroups (J1-P58)** Pierre A. Zalloua et al., "Y-Chromosomal Diversity in Lebanon Is Structured by Recent Historical Events," *American Journal of Human Genetics* 82, no. 4 (2008): 873–82.

101 **number of mutations** Badro et al., "Y-Chromosome and mtDNA Genetics."

102 **increasing social complexity** Klaus Schmidt, "Göbekli Tepe,

Southeastern Turkey: A Preliminary Report on the 1995–1999 Excavations," *Paléorient* 26, no. 1 (2000): 45–54.

CHAPTER 10: WHAT IS AN INDIGENOUS POPULATION?

105 **These people are called the Laal** Florian Lionnet, *Click Consonants* (Leiden: Brill, 2020), chap. 14.

110 **back-migration events to Africa** M. Haber et al., "Chad Genetic Diversity Reveals an African History Marked by Multiple Holocene Eurasian Migrations," *American Journal of Human Genetics* 99, no. 6 (2016): 1316–24.

110 **lush with vegetation** S. Kröpelin et al., "Climate-Driven Ecosystem Succession in the Sahara: The Past 6000 Years," *Science* 320, no. 5877 (2008): 765.

118 **contact with Eurasian farmers** M. G. Llorente et al., "Ancient Ethiopian Genome Reveals Extensive Eurasian Admixture in Eastern Africa," *Science* 350, no. 6262 (2015): 820.

CHAPTER 11: THE EARLY DYNASTIES AND EMPIRES

126 **"a watershed moment"** Jared Diamond, *Guns, Germs, and Steel: The Fates of Human Societies* (New York: Norton, 1997).

127 **our closest primate relatives, the chimps** Margaret Power, *The Egalitarians: Human and Chimpanzee* (Cambridge: Cambridge University Press, 1991).

130 **"Sumerian pictographic characters"** Charles K. Maisels, *The Emergence of Civilization: From Hunting and Gathering to Agriculture, Cities, and the State in the Near East* (London: Routledge, 2016); Harriet Crawford, *The Sumerian World* (New York: Routledge, 2013).

131 **Eblaite, another East Semitic language** D. D. Luckenbill, "Akkadian Origins," *American Journal of Semitic Languages and Literatures* 40, no. 1 (1923): 1–13.

131 **Akkadian culture arose** Crawford, *Sumerian World.*

132 **first Babylonian dynasty** Albert T. Clay, Review of *Amorites and Their Empire* by A. T. Olmstead, *American Journal of Theology* 23, no. 4 (1919): 525–27.

132 **claiming Babylon as its capital** Harvey W. Kish, "Akkad and Agade," *Journal of the American Oriental Society* 95, no. 3 (1975): 434–53.

136 **under assault by the Elamites** A. T. Olmstead, "The Babylonian Empire," *American Journal of Semitic Languages and Literatures* 35, no. 2 (1919): 65–100; Ada Taggar-Cohen, "The Kingdom of the Hittites: The Least Known Empire of the Second Millennium BCE," *Hebrew Studies* 52 (2011): 379–96.

CHAPTER 12: THE EARLY TRIBES

138 **The Medes were succeeded** R. Pleiner and J. K. Bjorkman, "The Assyrian Iron Age: The History of Iron in the Assyrian Civilization," *Proceedings of the American Philosophical Society* 118, no. 3 (1974): 283–313; A. R. Millard, "Assyrians and Arameans," *Iraq* 45, no. 1 (1983): 101.

140 **Maaloula, a mountain town** Millard, "Assyrians and Arameans," 101.

141 **the Land of Canaan** Sabatino Moscati, *Ancient Semitic Civilisations* (London: Elek Books, 1957).

142 **Philistines may have come from Crete** M. Feldman et al., "Ancient DNA Sheds Light on the Genetic Origins of Early Iron Age Philistines," *Science Advances* 5, no. 7 (2019): eaax0061.

142 **About fifty years later** Moscati, *Ancient Semitic Civilisations;* R. Zadok, "Phoenicians, Philistines, and Moabites in Mesopotamia," *Bulletin of the American Schools of Oriental Research* 230 (1978): 57–65.

143 **arrival of the Hyksos** Moscati, *Ancient Semitic Civilisations.*

144 **the Greek noun *Phoinix,* which means "purple"** Maria Eugenia Aubet, *The Phoenicians and the West: Politics, Colonies and Trade* (Cambridge: Cambridge University Press, 2001).

145 **establishment of Byblos** Sabatino Moscati and G. Palazzo, *The Phoenicians* (New York: Rizzoli, 1999).

145 **perfect maritime city** Sabatino Moscati, *The World of the Phoenicians* (London: Phoenix Giant, 1999); M. Dunand, *Byblos: son histoire, ses ruines ses legendes* (Beirut: n.p., 1973).

147 **active trading posts** Aubet, *Phoenicians and the West;* Maria Eugenia Aubet, *Commerce and Colonization in the Mediterranean Bronze Age* (Cambridge: Cambridge University Press, 2012).

148 **Sicily, Sardinia, and Malta** Moscati and Palazzo, *Phoenicians*; Moscati, *World of the Phoenicians*.

152 **Legend has it** R. T. Ridley, "To Be Taken with a Pinch of Salt: The Destruction of Carthage," *Classical Philology* 81, no. 2 (1986): 140–46.

152 **With the defeat of the Carthaginians** Aubet, *Phoenicians and the West;* Moscati and Palazzo, *Phoenicians;* Moscati, *World of the Phoenicians;* Aubet, *Commerce and Colonization*.

CHAPTER 13: THE PHOENICIANS

155 **"Most of the men"** Herodotus, *The Histories,* trans. Tom Holland, ed. Paul Cartledge (London: Penguin Classics, 2013).

165 **Phoenician signature is currently** Pierre A. Zalloua et al., "Identifying Genetic Traces of Historical Expansions: Phoenician Footprints in the Mediterranean," *American Journal of Human Genetics* 83, no. 5 (2008): 633–42.

166 **Phoenician burials at the Monte Sirai site** Elizabeth Matisoo-Smith et al., "Ancient Mitogenomes of Phoenicians from Sardinia and Lebanon: A Story of Settlement, Integration, and Female Mobility," *PLoS ONE* 13, no. 1 (2018): e0190169.

CHAPTER 14: THE FIRST ALPHABET

168 **"The supreme specimen"** Maurice Dunand, *Byblia grammata. Documents et recherches sur le développement de l'écriture en Phénicie* (Beirut, 1945), pl. 16.

169 **invented by the Sumerians** Piotr Michalowski, "Sumerian," in Rebecca Hasselbach-Andee, ed., *A Companion to Ancient Near Eastern Languages* (Hoboken, NJ: Wiley, 2020), 83–105.

169 **The Linear A script** Robert Hetzron, *The Semitic Languages* (London: Routledge, 2006); Roger D. Woodard, *The Ancient Languages of Syria-Palestine and Arabia* (Cambridge: Cambridge University Press, 2008).

170 **Pseudo-Hieroglyph script** Joaquim Azevedo, "The Origin and Transmission of the Alphabet," master's thesis, Andrews University, 1994; Ludwig Morenz, "The Development of Egyptian Writing in the Fourth and Early Third Millennium BCE," in Rebecca Hasselbach-Andee, ed., *A Companion to Ancient Near Eastern Languages* (Hoboken, NJ: Wiley, 2020), 47–64.

171 **Serabit El-Khadem inscriptions** Joseph Naveh, "Some Semitic Epigraphical Considerations on the Antiquity of the Greek Alphabet," *American Journal of Archaeology* 77, no. 1 (1973): 1.

171 **illiterate Canaanite settlers** Orly Goldwasser, "How the Alphabet Was Born from Hieroglyphs," *Biblical Archaeology Review* 36, no. 2 (2010): 36.

171 **"The alphabet was invented"** Anson F. Rainey, "Rainey's First Critique: Turquoise Miners Did Not Invent the Alphabet," Biblical Archaeology Society, *Bible History Daily,* August, 25, 2010.

173 **invented the Ugaritic cuneiform script** R. R. Stieglitz, "The Ugaritic Cuneiform and Canaanite Linear Alphabets," *Journal of Near Eastern Studies* 30, no. 2 (1971): 135–39; Christopher Rollston, "The Emergence of Alphabetic Scripts," in Rebecca Hasselbach-Andee, ed., *A Companion to Ancient Near Eastern Languages* (Hoboken, NJ: Wiley, 2020), 65–81.

173 **Three facts are** Azevedo, "Origin and Transmission of the Alphabet"; André Lemaire, "The Spread of Alphabetic Scripts (c. 1700–500 BCE)," *Diogenes* 55, no. 2 (2008): 45–58; Frank Moore Cross and Thomas O. Lambdin, "A Ugaritic Abecedary and the Origins of the Proto-Canaanite Alphabet," *Bulletin of the American Schools of Oriental Research* 160 (December 1960): 21–26; Christopher Woods, "The Emergence of Cuneiform Writing," in Rebecca Hasselbach-Andee, ed., *A Companion to Ancient Near Eastern Languages* (Hoboken, NJ: Wiley, 2020), 27–46.

173 **The scribes of Byblos** Françoise Briquel Chatonnet and Robert Hawley, "Phoenician and Punic," in Rebecca Hasselbach-Andee, ed., *A Companion to Ancient Near Eastern Languages* (Hoboken, NJ: Wiley, 2020), 297–318.

174 **simplified Egyptian signs with Canaanite values** John C. C. Clarke, "The Origin of the Semitic Alphabet," *Hebrew Student* 2, no. 10 (1883): 309–15; Frank Moore Cross, "The Evolution of the Proto-Canaanite Alphabet," *Bulletin of the American Schools of Oriental Research* 134 (1954): 15.

178 **"They transmitted"** Herodotus, *The Histories,* trans. Tom Holland, ed. Paul Cartledge (London: Penguin Classics, 2013).

179 **Philistines who adopted** Naveh, "Some Semitic Epigraphical Considerations on the Antiquity of the Greek Alphabet"; Joseph Naveh, "The Greek Alphabet: New Evidence," *Biblical Archaeologist* 43, no. 1 (1980): 22–25.

CHAPTER 15: A COMPLEX NARRATIVE

183 **the Levant consists** Christopher Woods, "The Emergence of Cuneiform Writing," in Rebecca Hasselbach-Andee, ed., *A Companion to Ancient Near Eastern Languages* (Hoboken, NJ: Wiley, 2020), 27–46.

185 **The current Jewish populations** Etan Levine, *Diaspora: Exile and the Contemporary Jewish Condition* (Tel Aviv: Shapolsky, 1986).

185 **Jewish genetics have** Harry Ostrer, "A Genetic Profile of Contemporary Jewish Populations," *Nature Reviews Genetics* 2 (2001): 891; Harry Ostrer and Karl Skorecki, "The Population Genetics of the Jewish People," *Human Genetics* 132, no. 2 (2013): 119–27; Doron M. Behar et al., "Counting the Founders: The Matrilineal Genetic Ancestry of the Jewish Diaspora," *PLoS ONE* 3, no. 4 (2008): e2062; Doron M. Behar et al., "The Genetic Variation in the R1a Clade Among the Ashkenazi Levites' Y Chromosome," *Scientific Reports* 7, no. 1 (2017): 14969.

186 **their impact alone is not substantial enough** Franck Salameh, *The Other Middle East: An Anthology of Modern Levantine Literature* (New Haven, CT: Yale University Press, 2017).

187 **Seleucus assumed power** André Lemaire, *Levantine Epigraphy and History in the Achaemenid Period (539–332 BCE)* (New York: Oxford University Press, 2016).

188 **he encouraged the settlement of Greeks** W. W. Tarn, *Alexander the Great,* vol. 2 (Cambridge: Cambridge University Press, 2014).

189 **the Sassanid Empire** Chester G. Starr and D.S.C. Rogers, *A History of the Ancient World,* 4th ed. (Oxford: Oxford University Press, 1991); Philip de Souza, *The Ancient World at War: A Global History* (London: Thames & Hudson, 2008).

190 **Berytian (modern-day Beirut) school of law** Elio Lo Cascio and Laurens Ernst Tacoma, eds., *The Impact of Mobility and Migration in the Roman Empire: Proceedings of the Twelfth Workshop of the International Network Impact of Empire, Rome, June 17–19, 2015* (Leiden: Brill, 2016); Tom Holland, *Rubicon: The Triumph and Tragedy of the Roman Republic* (London: Abacus, 2013).

191 **Most of the Roman soldiers in the Levant** Luuk de Ligt and Laurens Ernst Tacoma, *Migration and Mobility in the Early Roman Empire* (Leiden: Brill, 2016).

191 **Byzantines established many permanent movements** Elsa Marston, *The Byzantine Empire* (New York: Benchmark Books, 2002).

192 **The Sassanid Empire, drained** Touraj Daryaee, "Sasanian Persia (ca. 224–651 CE)," *Iranian Studies* 31, nos. 3-4 (1998): 431–61.

192 **Byzantines regained some** George Ostrogorsky, "The Byzantine Empire in the World of the Seventh Century," *Dumbarton Oaks Papers* 13 (1959): 1–21; Hamilton A. R. Gibb, "Arab-Byzantine Relations Under the Umayyad Caliphate," *Dumbarton Oaks Papers* 12 (1958): 219–33; Michael Bonner, *Arab-Byzantine Relations in Early Islamic Times* (Farnham, Surrey: Ashgate-Variorum, 2009).

192 **The Seljuks waged wars** A.C.S. Peacock, *The Great Seljuk Empire* (Edinburgh: Edinburgh University Press, 2015).

193 **European Christians responded** Thomas Asbridge, *The First Crusade: A New History* (New York: Oxford University Press, 2005).

194 **Crusaders were struggling** Thomas F. Madden, *The New Concise History of the Crusades* (New York: Barnes & Noble, 2007).

194 **from the Mongolian plains** Kate Raphael, *Muslim Fortresses in the Levant: Between Crusaders and Mongols* (New York: Routledge, 2011); Robert G. Hoyland, *In God's Path: The Arab Conquests and the Creation of an Islamic Empire* (New York: Oxford University Press, 2014).

195 **Ain Jalut** Bernard Lewis, "The Mongols, the Turks and the Muslim Polity," *Transactions of the Royal Historical Society* 18 (1968): 49–68.

195 **marked the Levant with hatred** Amalia Levanoni, "The Mamluk Conception of the Sultanate," *International Journal of Middle East Studies* 26, no. 3 (1994): 373–92.

195 **Ottomans dominated** A. J. Toynbee, "The Ottoman Empire in World History," *Proceedings of the American Philosophical Society* 99, no. 3 (1955): 119–26; Elton L. Daniel, Review of *Struggle for Domination in the Middle East: The Ottoman-Mamluk War 1485–91* by Shai Har-El, *Journal of Islamic Studies* 8, no. 1 (1997): 102–4.

CHAPTER 16: THE RELIGIONS THAT SHAPED THE MODERN LEVANT

197 **Lebanon is nearly the same size** W. W. Harris, *Lebanon: A History, 600–2011* (Oxford: Oxford University Press, 2015).

198 **predates the foundation of Christianity** Marc Haber et al., "Genome-Wide Diversity in the Levant Reveals Recent Structuring by Culture," *PLoS Genetics* 9, no. 2 (2013): e1003316.

201 **cannot discount reconversions** Ibid.

202 **fifty different populations** Luigi Luca Cavalli-Sforza, "The Human Genome Diversity Project: Past, Present and Future," *Nature Reviews Genetics* 6 (2005): 333.

203 **concentrated in the Chouf** Anis Obeid, *Druze and Their Faith in Tawhid* (Syracuse, NY: Syracuse University Press, 2006); Robert Brenton Betts, Review of *The Druzes: A New Study of Their History, Faith, and Society* by Nejla M. Abu-Izzedin, *International Journal of Middle East Studies* 21, no. 1 (1989): 148–50.

207 **Maronites, who live primarily** Butros Dau, *Religious, Cultural and Political History of the Maronites* (Lebanon: self-published, 1984).

208 **a group called the Mardaites** Matti Moosa, "The Relation of the Maronites of Lebanon to the Mardaites and Al-Jarājima," *Speculum* 44, no. 4 (1969): 597–608.

208 **Mu'awiyah had to pay a tax** Matti Moosa, *The Maronites in History* (Syracuse, NY: Syracuse University Press, 1986); Charles A. Kimball, Review of *The Maronites in History* by Matti Moosa, *Journal of Church and State* 30, no. 2 (1988): 351–52.

209 **20 percent in certain parts** Marc Haber et al., "Influences of History, Geography, and Religion on Genetic Structure: The Maronites in Lebanon," *European Journal of Human Genetics* 19, no. 3 (2011): 334–40.

211 **Approximately 8,200 years ago** Marianna Kulkova et al., "The 8200 calBP Climate Event and the Spread of the Neolithic in

Eastern Europe," *Documenta Praehistorica* 42 (December 2015); Eelco Rohling and Heiko Pälike, "Centennial-Scale Climate Cooling with a Sudden Cold Event Around 8,200 Years Ago," *Nature* 434 (2005): 975–79.

211 **The northern Lebanese mountains** Daniel E. Platt et al., "Autosomal Genetics and Y-Chromosome Haplogroup L1b-M317 Reveal Mount Lebanon Maronites as a Persistently Non-Emigrating Population," *European Journal of Human Genetics* 29, no. 4 (2020): 581–92.

212 **The Libyan L1b** Ibid.

212 **key role in exchanging** Aurélien Girard and Giovanni Pizzorusso, "The Maronite College in Early Modern Rome: Between the Ottoman Empire and the Republic of Letters," in Liam Chambers and Thomas O'Connor, eds., *College Communities Abroad: Education, Migration and Catholicism in Early Modern Europe* (Oxford: Oxford University Press, 2017).

ABOUT THE AUTHOR

PIERRE ZALLOUA is a population geneticist focusing primarily on eastern Mediterranean populations. He earned his PhD in genetics from the University of California, Davis. He has held academic positions at Harvard University, Khalifa University, the American University of Beirut, Lebanese American University, and the University of Balamand, and has authored and co-authored more than 180 peer-reviewed manuscripts. Zalloua is featured in a documentary film about his work produced by National Geographic entitled *Quest for the Phoenicians*.

ABOUT THE TYPE

This book was set in Bembo, a typeface based on an old-style Roman face that was used for Cardinal Pietro Bembo's tract *De Aetna* in 1495. Bembo was cut by Francesco Griffo (1450–1518) in the early sixteenth century for Italian Renaissance printer and publisher Aldus Manutius (1449–1515). The Lanston Monotype Company of Philadelphia brought the well-proportioned letterforms of Bembo to the United States in the 1930s.